Applications of Group Theory in Quantum Mechanics

M. I. Petrashen & E. D. Trifonov

Translated by S. Chomet
Edited and Revised by J. L. Martin

Dover Publications, Inc.
Mineola, New York

Copyright

Bibliographical Note

This Dover edition, first published in 2009, is an unabridged republication of the work first published in English in 1969 by The M.I.T. Press, Cambridge, Massachusetts. It was originally published in Moscow under the title *Primeneniye Teorii Grupp v Kvantovoi Mekhanike.*

Library of Congress Cataloging-in-Publication Data

Petrashen, M. I. (Mariia Ivanovna)
[Primenenie teorii grupp v kvantovoi mekhanike. English]
Applications of group theory in quantum mechanics / M. I. Petrashen and E.D. Trifonov. — Dover ed.
p. cm.
Originally published: Cambridge, Mass. : M.I.T., 1969.
Includes bibliographical references and index.
ISBN-13: 978-0-486-47223-2
ISBN-10: 0-486-47223-X
1. Group theory. 2. Quantum theory. I. Trifonov, E. D. (Evgenii Dmitrievich), joint author. II. Title.

QA174.2.P4813 2008
530.1201'5122—dc22

2008044542

Manufactured in the United States of America
Dover Publications, Inc., 31 East 2nd Street, Mineola, N.Y. 11501

Contents

Foreword

This monograph is based on a course of lectures on the applications of group theory to problems in quantum mechanics, given by the authors to undergraduates at the Physics Department of Leningrad University.

Following a period of scepticism about the value of group theory as a means of investigating physical systems, this mathematical theory eventually won a very general acceptance by physicists. The group-theory formalism is now widely used in various branches of quantum physics, including the theory of the atom, the theory of the solid state, quantum chemistry, and so on. Recent achievements in the theory of elementary particles, which are intimately connected with the application of group theory, have intensified general interest in the possibility of using group-theoretical methods in physics, and have shown once again the importance and eminent suitability of such methods in quantum theory.

A relatively large number of textbooks and monographs on applications of group theory in physics is already available. A bibliography is given at the end of the book.

The range of applications of the methods of group

theory to physics is continually expanding, and it is hardly possible at the present time to produce a monograph which would cover all these applications. The best course to adopt, therefore, is to include the relevant applications in monographs or textbooks devoted to special topics in physics. This is done, for example, in the well-known course on theoretical physics by Landau and Lifshits. It is likely that this tendency will continue in the future.

At the same time, a theoretical physicist should have a general knowledge of the leading ideas and methods of group theory as used in physics. Our aim in this course was to satisfy this need. Moreover, we thought it would be useful to include in the book a number of problems which have not been discussed in existing monographs, or treated in sufficient detail. We refer, above all, to studies of the symmetry properties of the Schroedinger wave function, to the explanation of 'additional' degeneracy in the Coulomb field, and to certain problems in solid-state physics.

In our course, we have restricted our attention to applications of group theory to quantum mechanics. It follows that the book can be regarded as the first part of a broader course, the second part of which should be devoted to applications of group-theoretical methods to quantum field theory. We conclude our book with an account of related problems concerned with the conditions for relativistic invariance in quantum theory.

We are grateful to M. N. Adamov, who read this monograph in manuscript and made a number of valuable suggestions, and to A. G. Zhilich and I. B. Levinson, who reviewed individual chapters. In the preparation of the manuscript for press we made use of the kind assistance of A. A. Kiselev, B. Ya. Frezinskii, R. A. Evarestov, A. A. Berezin and G. A. Natanzon.

Chapter 1

Introduction

In the first chapter of this monograph we shall try, in so far as it is possible at the beginning of a book, to show how one can naturally and advantageously apply the theory of groups to the solution of physical problems. We hope that this will help the reader who is mainly interested in the applications of group theory to physics to become familiar with the general ideas of abstract groups which are necessary for applications.

1.1 Symmetry properties of physical systems

It is frequently possible to establish the properties of physical systems in the form of symmetry laws. These laws are expressed by the invariance (invariant form) of the equations of motion under certain definite transformations. If, for example, the equations of motion are invariant under orthogonal transformations of Cartesian coordinates in three-dimensional space, it may be concluded that reference frames oriented in a definite way relative to each other are equivalent for the description of the motion of the physical system under consideration. Equivalent reference frames are usually defined as frames in which identical phenomena occur in the same way when identical initial conditions are set up for them. Conversely, if in a physical theory it is postulated that certain reference frames are equivalent, then the equations of motion should be invariant under the transformations relating the coordinates in these systems. For example, the postulate of the theory of relativity which demands the equivalence of all reference frames moving with uniform velocity relative to one another is expressed by the invariance of the equations of motion under the Lorentz transformation. The class of equivalent reference frames for a given problem is frequently determined from simple geometrical considerations applied to a model of the physical system. This is done, for example, in the case of symmetric molecules, crystals and so on. However, not all transformations under which the equations of motion are invariant can be interpreted as transformations to a new reference frame. The symmetry of a physical system may not have an immediate geometrical interpretation. For example, V. A. Fock has shown that the Schroedinger equation for the hydrogen atom is invariant under rotations in a four-dimensional space connected with the momentum space.

The symmetry properties of a physical system are general and very important features. Their generality usually ensures that they remain valid while our knowledge of a given physical system grows. They must not, however, be regarded as absolute properties; like any other descriptions of physical systems they are essentially approximate. The approximate nature of some symmetry properties is connected with the current state of our knowledge, while in other cases it is due to the use of simplified models of physical systems which facilitate the solution of practical problems.

Thus, by the symmetry of a system we shall not always understand the invariance of its equations of motion under a certain set of transformations. The following important property must always be remembered: if an equation is invariant under transformations A and B, it is also invariant under a transformation C which is the result of the successive application of the transformations A and B. The transformation C is usually called the product of the transformations A and B. A set of symmetry transformations for a given physical system is therefore closed with respect to the operation of multiplication which we have just defined. Such a set of transformations is called a group of symmetry transformations for the given physical system. A rigorous definition of a group is given below.

1.2 Definition of a group

A group G is defined as a set of objects or operations (elements of the group) having the following properties.

1. The set is subject to a definite 'multiplication' rule, i.e. a rule by which to any two elements A and B of the set G, taken in a definite order, there corresponds a unique element C of this set which is called the product of A and B. The product is written $C = AB$.

2. The product is associative, i.e. the equation $(AB)D = A(BD)$ is satisfied by any elements A, B and D of the set. The product may not be commutative, i.e. in general $AB \neq BA$. Groups for which multiplication is commutative are Abelian.

3. The set contains a unique element E (the identity or unit element) such that the equation

$$AE = EA = A$$

is satisfied by any element A in the set.

4. The set G always includes an element F (the inverse) such that for any element A

$$AF = E$$

The inverse is usually denoted by A^{-1}.

The above four properties define a group. We see that a group is a set which is closed with respect to the given rule of multiplication. The following are consequences of the above properties.

a. The group contains only one unit element. Thus, for example, if we suppose that there are two unit elements E and E' in the group G, then in view of property 3 we have

$$EE' = E = E'E = E'$$

i.e. $E = E'$.

b. If F is the inverse of A, the element A will be the inverse of F, i.e. if $AF = E$, then $FA = E$. In fact, multiplying the first of these equations on the left by F, we have

$$FAF = F$$

The element F (like any other element of the set G) has an inverse F^{-1}. Multiplying the last equation on the right by F^{-1} we obtain $FAFF^{-1} = FF^{-1}$, i.e. $FA = E$.

c. For each element in the set there is only one inverse element. Let us suppose that an element A in G has two inverse elements F and D, i.e. $AF = E$ and $AD = E$. If this is so, then by multiplying the equation $AF = AD$ on the left by A^{-1} we obtain $F = D$.

d. If $C = AB$ then $C^{-1} = B^{-1}A^{-1}$, because of the associative property of the product of two elements in the group.

We note also that if the number of elements in a group is finite, then the group is called a finite group; if the number of elements is infinite, the group is called an infinite group. The number of elements in a finite group is the order of the group.

The following are examples of groups.

1. The set of all integers, including zero, forms an infinite group if addition is taken as group multiplication. The unit element in this group is 0, the inverse element of a number A is $-A$, and the group is clearly Abelian.

2. The set of all rational numbers, excluding zero,

forms a group for which the multiplication rule is the same as the familiar multiplication rule used in arithmetic. The unit element is 1. This is again an infinite Abelian group. The positive rational numbers also form a group, but the negative rational numbers do not.

3. The set of vectors in n-dimensional linear space forms a group. The group multiplication rule is the vector addition; the unit element is the zero vector and the inverse of a vector $\boldsymbol{a}$ is $-\boldsymbol{a}$.

4. The set of all non-singular n-th order matrices (or the corresponding linear transformations in n-dimensional space), $GL(n)$, is an example of a non-Abelian group. It is clear that the elements of this group depend on n^2 continuously varying parameters (elements of the group). Infinite groups whose elements depend on continuously varying parameters are continuous groups. The unit element of the group $GL(n)$ is the unit matrix; the inverse elements are the corresponding inverse matrices. The operation of group multiplication is the same as the rule of multiplication of matrices, which is not commutative.

1.3 Examples of groups used in physics

Let us now list some groups which will be used in applications.

1. The three-dimensional translation group. The elements of this group are the displacements of the origin of coordinates through an arbitrary vector $\boldsymbol{a}$:

$$\boldsymbol{r}' = \boldsymbol{r} + \boldsymbol{a}$$

It is clear that this is a three-parameter (three components of the vector $\boldsymbol{a}$) continuous group.

2. The rotation group $O^+(3)$. The elements of this group are rotations of three-dimensional space, or the corresponding orthogonal matrices with a determinant equal to unity. This is also a continuous three-parameter group: the

nine elements of the orthogonal transformation matrix are related by six conditions, and three angles $\{\varphi, \theta, \psi\}$ can be taken as the independent rotation parameters. The polar angles φ and θ define the position of the rotational axis passing through the origin, and the angle ψ defines rotation about this axis (see Exercise 1.1). Invariance with respect to the group $O^+(3)$ expresses the isotropy of three-dimensional space, i.e. the equivalence of all directions in this space.

If we add the operations of rotation accompanied by inversion (e.g. $x' = -x,\ y' = -y,\ z' = -z$) to the rotation group we obtain the orthogonal group $O(3)$.

3. Molecular symmetry groups, i.e. point groups, consist of certain orthogonal transformations of three-dimensional space. For example, the symmetry group of a molecule having the configuration of an octahedron consists of 48 elements, namely, rotations and rotations accompanied by inversion which transform the corners of a cube into one another.

4. The crystal symmetry groups, or space groups, consist of a finite number of orthogonal transformations and discrete translations, and all products of these transformations. Strictly speaking, such symmetry is exhibited only by an infinite crystal or a model of a crystal with the so-called periodic boundary conditions.

5. The permutation group which consists of all permutations of n symbols, e.g. the coordinates of n identical objects. This is a finite group of order $n!$.

6. The Lorentz group L^+ consists of transformations relating the coordinates of two reference frames which are in uniform rectilinear relative motion. This group includes the rotation group $O^+(3)$ and depends on six parameters, namely, three angles defining the mutual orientation of the space axes, and the three components of the relative velocity. The invariance of the equations of motion under the Lorentz group is a consequence of the postulates of the theory of relativity.

The groups listed above do not, of course, exhaust all the possibilities as far as applications in physics are concerned. We shall, however, devote most of our attention to the above groups.

1.4 Invariance of equations of motion

We shall now consider the invariance of the equations of motion of a physical system with respect to transformations of its symmetry group.

In classical mechanics the motion of a system is described by Lagrange's equations. The symmetry of a physical system with respect to a given transformation group is therefore expressed through the invariance of Lagrange's equations (and additional conditions, if such exist) with respect to these transformations. Since the equations of motion written in terms of the Lagrangian $\mathscr{L}$ for any chosen generalized coordinates q_i are always of the same form, i.e.

$$\frac{d}{dt}\frac{\partial \mathscr{L}}{\partial \dot{q}_i} - \frac{\partial \mathscr{L}}{\partial q_i} = 0 \tag{1.1}$$

it follows that their invariance will be ensured if the Lagrangian itself is invariant. It is important to note, however, that the requirement that the Lagrangian should be invariant is too stringent. We know that the equations of motion remain unaltered when the Lagrangian is multiplied by a number, and a time derivative of an arbitrary function of the generalized coordinates is added to it. For example, the symmetry of the one-dimensional harmonic oscillator with respect to the interchange of coordinates and momenta (a so-called content transformation in classical mechanics) corresponds to a change of the sign of its Lagrangian

$$\mathscr{L} = \frac{1}{2}p^2 - \frac{1}{2}q^2$$

In quantum mechanics the state of a physical system

is described by a wave function $\psi(x, t)$, which is the solution of the Schroedinger equation

$$\hat{H}(x)\psi(x, t) = i\hbar\frac{\partial}{\partial t}\psi(x, t) \tag{1.2}$$

The symmetry of a quantum-mechanical system with respect to a given group is therefore reflected in the invariance of the Schroedinger equation under the transformations in this group. If the symmetry group consists of transformations of the configuration space

$$x' = ux$$

then the invariance of the Schroedinger equation can be verified by substituting

$$x = u^{-1}x', \quad \psi'(x') = \psi(u^{-1}x') \tag{1.3}$$

If the Schroedinger equation is invariant under the transformation u, then it should retain its form after the substitution of (1.3) in (1.2). It is clear that this will be so if the substitution does not alter the form of the Hamiltonian $\hat{H}(x)$.

Group theory enables us to classify the states of a physical system entirely on the basis of its symmetry properties and without carrying out an explicit solution of the equations of motion. This is, in fact, the basic value of the group-theoretical method, since even an approximate solution of the equations of motion is frequently very difficult. By applying group-theoretical methods we can establish the symmetry properties of the exact solutions of these equations, and thus deduce important information about the physical system under consideration.

Although we are not yet ready to use the group-theory formalism we shall, nevertheless, try to illustrate these ideas by taking an example from classical mechanics. We know that in classical mechanics the classification of the motions of a given system is based on the values of its integrals (constants) of motion. We shall show that the existence of these integrals is due to the symmetry of the system with respect to a group of continuous transformations. Consider a system of mass points for which the Lagrangian is invariant under the translation group in three-dimensional space. This means that the change in the Lagrangian due to the translation

$$\mathbf{r}_i' = \mathbf{r}_i + \mathbf{a}, \quad \delta\mathbf{r}_i = \mathbf{a} \tag{1.4}$$

must be zero. Assuming that $\mathbf{a}$ is an infinitesimal vector, we have

$$\delta\mathcal{L} = \sum_i \frac{\partial\mathcal{L}}{\partial\mathbf{r}_i}\,\delta\mathbf{r}_i = \mathbf{a}\sum_i \frac{\partial\mathcal{L}}{\partial\mathbf{r}_i} = 0 \tag{1.5}$$

Using the Lagrange equations

$$\frac{\partial\mathcal{L}}{\partial\mathbf{r}_i} = \frac{d}{dt}\frac{\partial\mathcal{L}}{\partial\dot{\mathbf{r}}_i} \tag{1.6}$$

and the fact that $\mathbf{a}$ is arbitrary, we have

$$\frac{d}{dt}\sum_i \frac{\partial\mathcal{L}}{\partial\dot{\mathbf{r}}_i} = 0 \tag{1.7}$$

or

$$\sum_i \frac{\partial\mathcal{L}}{\partial\dot{\mathbf{r}}_i} \equiv \sum_i \mathbf{p}_i = \mathbf{P} = \text{const} \tag{1.8}$$

Thus, from the invariance of the Lagrangian with respect to translations in three-dimensional space we deduce that the total momentum of the system is a constant of motion.

It can similarly be shown that the requirement of invariance under time translations ensures that the energy of the system is a constant of motion.

We shall show later that analogous results are valid in quantum mechanics.

Exercises

1.1. Show that any rotational transformation of three-dimensional space may be represented as a rotation through a definite angle about an axis passing through the origin.

1.2. Show that the invariance of the Lagrangian under the three-dimensional rotation group ensures that the total angular momentum of the system is a constant of motion.

Chapter 2

Abstract Groups

When we investigate the general properties of a group we need not specify the realization of its elements (by transformations, matrices, etc.). By denoting the elements of a group by certain symbols which obey a given rule of multiplication, we obtain the so-called abstract group. In this chapter we shall review some of the properties of such groups.

2.1 Translation along a group

Suppose that the group G consists of m elements $g_1, g_2, \ldots, g_m$. Let us multiply each element on the right by the same element g_l, i.e. let us carry out a right translation along the group. We thus obtain the sequence

$$g_1 g_l, \; g_2 g_l, \; \ldots, \; g_m g_l \tag{2.1}$$

We shall show that each group element is encountered once and only once in this sequence. In fact, let g_t be an arbitrary element of the group. It is clear that $g_t = (g_t g_l^{-1}) g_l$, and, consequently, the element g_t appears in the sequence (2.1).

Since the number of elements in our sequence is equal to the order of the group, each of the elements can be found in the sequence only once. The sequence of elements

$$g_i g_1,\ g_i g_2,\ \ldots,\ g_i g_m \tag{2.2}$$

which is obtained as a result of a left translation has the same property.

2.2 Sub-groups

A set of elements belonging to a group G, which itself forms a group with the same multiplication rule, is a sub-group of G. The remainder of the group G cannot form a group since, for example, it does not contain the unit element.

2.3 The order of an element

Let us take an arbitrary element g_l of the group G and consider the powers g_l, g_l^2, g_l^3, ... of this element. Since we are considering a finite group, the members of this sequence must appear repeatedly. Suppose, for example, that

$$g_l^{k_1} = g_l^{k_2} = g_l, \quad k_2 > k_1$$

We then have

$$g_l^{k_2} = g_l^{k_1} g_l^{k_2-k_1} = g_l g_l^{k_2-k_1} = g_l$$

and, consequently,

$$g_l^{k_2-k_1} = E$$

The smallest exponent h for which

$$g_l^h = E$$

is the order of the element g_l. The set of elements g_l, g_l^2, ..., $g_l^h = E$ is the period or cycle of the element g_l. It is clear that the period of an element forms a sub-group of G.

It is readily seen that all the elements of this sub-group commute and, consequently, the sub-group is Abelian.

If h is the order of the element g_i, then $g_i^{h-1} = g_i^{-1}$. Therefore, for finite groups, the existence of inverse elements is a consequence of the three other group properties.

2.4 Cosets

Let H be a sub-group of a group G with elements $h_1, h_2, \ldots, h_m$, where m is the order of H. Let us construct the following sequences of sets of elements of G. Let us first take from G an element g_1, which is not contained in H, and construct the set $g_1h_1, g_1h_2, \ldots, g_1h_m$, which we shall denote by g_1H. Next, let us take from G an element g_2, which is not contained in H or in g_1H, and set up the further set g_2H. We can continue this process until we exhaust the entire group. As a result, we obtain the sequence

$$H, \ g_1H, \ g_2H, \ \ldots, \ g_{k-1}H \tag{2.3}$$

The sets g_iH are the left cosets of the sub-group H.

We shall show that the cosets defined above have no common elements. In fact, let us suppose that the sets g_1H and g_2H have one common element: for example, $g_1h_1 = g_2h_2$. We then have $g_2 = g_1h_1h_2^{-1} = g_1h_3$, so that g_2 belongs to the set g_1H. This result, however, conflicts with our original assumption and, therefore, each element of the group G enters only one of the cosets. Since G contains n elements, and each of the cosets contains m elements, it follows that $m = \frac{n}{k}$. The number k is the index of the sub-group H in the group G. We thus see that the order of the sub-group is a divisor of the order of the group.

Similarly, we can decompose the group G into the right cosets

$$H, \ Hg'_1, \ Hg'_2, \ \ldots, \ Hg'_{k-1} \tag{2.4}$$

In constructing the cosets we have a choice in selecting

the element g_i. We shall show that for any acceptable choice of the elements g_i we obtain the same set of cosets and, consequently, the same decomposition. This result follows directly from the following theorem: two cosets g_iH and g_kH (g_i and g_k are any two elements of the group G) either coincide or have no common elements. In fact, if these sets have at least one common element $g_ih_\alpha = g_kh_\beta$, then $g_k = g_ih_\alpha h_\beta^{-1}$ and, consequently, $g_k \in g_iH$. However, any element of the set g_kH can then be represented in the form $g_kh_\gamma = g_ih_\alpha h_\beta^{-1}h_\gamma = g_ih_\delta$ and will also belong to the conjugate set g_iH.

The group G can therefore be uniquely decomposed into left (or right) cosets of the sub-group H.

2.5 Conjugate elements and class

Let g be an element of the group G and let us construct the element $g' = g_igg_i^{-1}$; $g_i \in G$. The elements g and g' are said to be conjugate. Let us suppose now that g_i runs over all the elements of the group G. We then obtain n elements, some of which may be equal. Let the number of distinct elements be k, and let us denote them by $g_1, g_2, \ldots, g_k$. It is clear that this set includes all the elements of the group G which are conjugate to the element g. Moreover, it is readily shown that all the elements of this set are mutually conjugate. In fact, let $g_1 = g_\alpha g g_\alpha^{-1}$, $g_2 = g_\beta g g_\beta^{-1}$. We then have $g = g_\alpha^{-1}g_1g_\alpha$ and $g_2 = g_\beta g_\alpha^{-1}g_1g_\alpha g_\beta^{-1} = g_\beta g_\alpha^{-1}g_1(g_\beta g_\alpha^{-1})^{-1}$. The set of all the mutually conjugate elements forms a class. Thus, the elements $g_1, g_2, \ldots, g_k$ form a class of conjugate elements. We see that the class is fully defined by specifying one of the elements. The number of elements in a class is its order. Any finite group can be divided into a number of classes of conjugate elements. The unit element of a group by itself forms a class. It is readily verified that all the elements of a given class have the same order.

We shall show that the set of products of the elements

of two classes consists of whole classes. This can be written as follows:

$$C_i C_j = \sum_k h_{ijk} C_k \tag{2.5}$$

where C_i is the set of elements of class i and h_{ijk} are integers. We shall first show that if $g_p \in C_i C_j$, then the entire class C_p to which g_p belongs itself belongs to the set $C_i C_j$. In fact, let $g_p = g_i g_j$, $g_i \in C_i$, $g_j \in C_j$. We then have for any $g \in G$

$$g^{-1} g_p g = g^{-1} g_i g g^{-1} g_j g \in C_i C_j \tag{2.6}$$

To prove (2.5) it remains to show that each element of the class C_p enters the set $C_i C_j$ the same number of times. Suppose, for example, that the element g_p enters twice, i.e.

$$g_i g_j = g_p \text{ and } g_{i'} g_{j'} = g_p \tag{2.7}$$

where

$$g_i \neq g_{i'}, \qquad g_j \neq g_{j'} \tag{2.8}$$

Each element $g'^{-1} g_p g'$ $(g' \in G)$ will then be contained in $C_i C_j$ at least twice. In fact,

$$\left.\begin{aligned} g'^{-1} g_p g' &= g'^{-1} g_i g_j g' = (g'^{-1} g_i g')(g'^{-1} g_j g') \\ g'^{-1} g_p g' &= g'^{-1} g_{i'} g_{j'} g' = (g'^{-1} g_{i'} g')(g'^{-1} g_{j'} g') \end{aligned}\right\} \tag{2.9}$$

and it follows from (2.8) that

$$g'^{-1} g_i g' \neq g'^{-1} g_{i'} g' \text{ and } g'^{-1} g_j g' \neq g'^{-1} g_{j'} g' \tag{2.10}$$

It is clear that the element $g'^{-1} g_p g'$ will not be encountered more than twice, since otherwise it can be shown by a similar argument that the element g_p is also encountered more than twice, which contradicts the original assumption.

2.6 Invariant sub-group (normal divisor)

Let H be a sub-group of the group G, and suppose that $g_i \in G$. Consider the set of elements $g_i H g_i^{-1}$, where g_i is fixed. This set is also a group, since all the group axioms are satisfied for it. Such a group is said to be similar to the sub-group H. If $g_i \in H$, then the similar sub-group will, of course, coincide with H. If, however, $g_i \notin H$, then in general we obtain a sub-group of G which is different from H. When the sub-group H coincides with all its similar sub-groups, it is called an invariant sub-group or a normal divisor. An invariant sub-group will be represented by the letter N. It follows from the definition that if an invariant sub-group contains an element g of the group G, then it will also contain the entire class to which g belongs. The invariant sub-group may thus be said to consist of whole classes of the group.

For an invariant sub-group of the group G, the left and right cosets coincide. In fact,

$$g_i N = g_i N g_i^{-1} g_i = N g_i \tag{2.11}$$

since

$$g_i N g_i^{-1} = N \tag{2.12}$$

Any group has two trivial invariant sub-groups: the first of these coincides with the group itself and the second consists of the unit element. Groups which do not have invariant sub-groups other than the trivial groups are called simple.

2.7 The factor group

Let N be an invariant sub-group of the group G. Let us decompose G into the cosets N:

$$N,\ g_1 N,\ g_2 N,\ \ldots,\ g_{k-1} N$$

and form a set $g_1 N g_2 N$, which consists of different elements $g_1 n_\alpha g_2 n_\beta$, where n_α and n_β run independently over the entire sub-group N. It is readily seen that

$$g_1 N g_2 N = g_1 g_2 g_2^{-1} N g_2 N = g_1 g_2 N N = g_1 g_2 N = g_3 N \tag{2.13}$$

If the set g_1Ng_2N is called the product of the sets g_1N and g_2N, then the product of two cosets of N will again give a coset of N. Next, multiplication (in the sense just indicated) of a coset of N by N on the left or right does not change this coset:

$$Ng_1N = g_1g_1^{-1}Ng_1N = g_1NN = g_1N \tag{2.14}$$

For each coset g_iN there is a coset $g_i^{-1}N$ such that their product is equal to N:

$$g_i^{-1}Ng_iN = NN = N \tag{2.15}$$

It follows from these results that the cosets of an invariant sub-group can be regarded as the elements of a new group in which N plays the role of the unit element. This group is the factor group or quotient group of the invariant sub-group. Its order is equal to the index of the invariant sub-group.

2.8 Isomorphism and homomorphism of groups

If between the elements of two groups there is a one-to-one correspondence which preserves group multiplication, then the groups are isomorphic. Thus, let G and $\tilde{G}$ be two isomorphic groups. Then if the elements g_i and g_k of G correspond to the elements $\tilde{g}_i$ and $\tilde{g}_k$ of $\tilde{G}$, i.e.

$$g_i \leftrightarrow \tilde{g}_i \qquad g_k \leftrightarrow \tilde{g}_k$$

then

$$g_ig_k = g_l \leftrightarrow \tilde{g}_l = \tilde{g}_i\tilde{g}_k$$

By establishing the isomorphism of groups we can reduce the investigation of a given group to that of another group isomorphic to it.

Another important concept in group theory is that of homomorphism. If to each element of a group G there corresponds only one definite element of the group $\tilde{G}$ and to each element of $\tilde{G}$ there corresponds a number of elements of G and, moreover, this correspondence is preserved under group multiplication, then the group $\tilde{G}$ is homomorphic to G. Homomorphism has the following properties.

a. If the group $\tilde{G}$ is homomorphic to G, then the unit element of G corresponds to the unit element of $\tilde{G}$. In fact, let E be the unit element of G, in which case for any $g \in G$ we have $Eg = gE = g$. Let E and $\tilde{g}$ be the elements of the group $\tilde{G}$ corresponding to E and g, in which case, since the groups are homomorphic, we have $\tilde{E}\tilde{g} = \tilde{g}\tilde{E} = \tilde{g}$. Hence it follows that $\tilde{E}$ is the unit element of $\tilde{G}$.

b. If the group $\tilde{G}$ is homomorphic to G, then mutually reciprocal elements of G correspond to mutually reciprocal elements of $\tilde{G}$. In fact, let $g_i g_k = E$. Then, in view of the correspondence, $\tilde{g}_i \tilde{g}_k = \tilde{E}$.

c. If the group $\tilde{G}$ is homomorphic to G, then all the elements of G which correspond to the unit element of $\tilde{G}$ form an invariant sub-group N of the group G. In fact, suppose that the elements $g'_1, g'_2, \ldots, g'_s$ of G correspond to the unit element $\tilde{E}$ of the group G. The product $g'_i g'_k$ then corresponds to $\tilde{E}\tilde{E} = \tilde{E}$. Consequently, $g'_i g'_k = g'_l$, and the set $g'_1, g'_2, \ldots, g'_s$ is closed with respect to group multiplication. According to property (a) it should contain the unit element, but since the unit element $\tilde{E}$ is the inverse of itself, it follows that, because of property (b), for each element g'_i we can find an inverse element g'_k. Next, from the equation $\tilde{g}\tilde{E}\tilde{g}^{-1} = \tilde{E}$, where $\tilde{g}$ is an arbitrary element of the group $\tilde{G}$, it follows that $g g'_i g^{-1} = g'_f$ for an arbitrary element g of the group G. The properties which we have established for the set $g'_1, g'_2, \ldots, g'_s$ are sufficient to enable us to conclude that this set forms an invariant sub-group of the group G.

d. If the group $\tilde{G}$ is homomorphic to G, the elements of G corresponding to the element $\tilde{g}_i$ form the conjugate set Ng_i, where g_i is any of the elements of G corresponding to the element $\tilde{g}_i$, and N is the invariant sub-group corresponding to the unit element of $\tilde{G}$.

To prove this property let us split the group G into the conjugate sets

$$N,\ g_1N,\ g_2N,\ \ldots,\ g_{k-1}N$$

To any element of the set $g_i N$ there corresponds the element $\tilde{g}_i \tilde{E} = \tilde{g}_i$, i.e. the same element $\tilde{g}_i$ of the group $\tilde{G}$. It remains to show that different elements correspond to different conjugate sets. Let us assume that the opposite is the case. Thus, suppose that the element $\tilde{g}_1$ of the group $\tilde{G}$ corresponds to the sets $g_1 N$ and $g_2 N$. The element $\tilde{g}_1^{-1}\tilde{g}_1 = \tilde{E}$ then corresponds to the element $g_1^{-1}g_2$, and hence it follows that $g_1^{-1}g_2$ belongs to N. But then $g_1^{-1}g_2 = g'_k$ and $g_2 = g_1 g'_k$, which contradicts the original assumption that the sets $g_1 N$ and $g_2 N$ are different. It follows that there is a one-to-one correspondence between the conjugate sets $g_i N$ and the elements of the group $\tilde{G}$. It follows that the group $\tilde{G}$ is isomorphic to the factor group of the invariant sub-group N in G.

With this we conclude our review of the general properties of finite groups. A number of special theorems will be proved later in connection with the applications of the methods of group theory to physical problems.

Exercises

2. 1. The elements E, A, B, C, D, F form the group S_6 of order 6 with the following multiplication table (the first factors are shown in the first column, for example, $AB = D$):

	E	A	B	C	D	F
E	E	A	B	C	D	F
A	A	E	D	F	B	C
B	B	F	E	D	C	A
C	C	D	F	E	A	B
D	D	C	A	B	F	E
F	F	B	C	A	E	D

a. Find the orders of all the elements.

b. Find the sub-groups.

c. Divide the group into cosets and verify that this can be done in a unique way.

d. Divide the group into classes of conjugate elements.

e. Find the invariant sub-groups and verify that the right and left cosets are the same for each invariant sub-group.

f. Write down the multiplication tables for the corresponding factor groups.

g. Show that the abstract group S_6 has the following realizations: permutation group of three elements, and matrix group of order 2 corresponding to rotations and reflections in a plane which transform the apices of an equilateral triangle into one another.

2. 2. Show that the order of a group is equal to the order of any of its elements multiplied by an integer.

2. 3. Using the concept of the order of a group element, construct the multiplication tables for the possible groups of order 3 and 4.

2. 4. Show that all the elements of a given class have the same order.

2. 5. Show that any sub-group of index 2 is invariant.

2. 6. Show that in the set gg_ig^{-1}, where g runs over the group, each element of the class to which g_i belongs is encountered the same number of times.

Chapter 3

Representations of Point Groups

3.1 Definition of a representation of a group

Consider a finite group G with elements $g_1, g_2, \ldots, g_m$. If a group T of linear operators $\hat{T}_{g_i}$ in a space R is homomorphic to G, then the group T is said to form a representation of G. Homomorphism leads to

$$\hat{T}_{g_i}\hat{T}_{g_k} = \hat{T}_{g_i g_k} \tag{3.1}$$

If the space R is the n-dimensional vector space R_n, then any of its elements $\boldsymbol{x}$ can be expanded in terms of n unit vectors $\boldsymbol{e}_k$ forming the basis of this space:

$$x = x_1 e_1 + x_2 e_2 + \ldots + x_n \boldsymbol{e}_n$$

The operator $\hat{T}_{g_i}$ will be defined if we specify its effect on each of the unit vectors $\boldsymbol{e}_k$. Suppose that

$$\hat{T}_{g_i} \boldsymbol{e}_k = \sum_{r=1}^{n} D_{rk}(g_i)\, \boldsymbol{e}_r \tag{3.2}$$

It is clear that to each element g_i of our group we can assign a matrix $\| D_{rk}(g_i) \|$. It is also clear that the unit element of the group can be associated with a unit matrix, and the inverse elements can be associated with inverse matrices. Let us show that for the matrices D we have

$$D(g_i)\,D(g_j) = D(g_i g_j) \tag{3.3}$$

In fact, if we apply the operators $\hat{T}_{g_j}$ and $\hat{T}_{g_i}$ successively to the unit vector e_k, we obtain

$$\hat{T}_{g_i}\hat{T}_{g_j}e_k = \hat{T}_{g_i}\sum_r D_{rk}(g_j)\,e_r$$

$$= \sum_{f,r} D_{rk}(g_j)\,D_{fr}(g_i)\,e_f = \sum_f \left(\sum_r D_{fr}(g_i)\,D_{rk}(g_j)\right) e_f \tag{3.4}$$

On the other hand,

$$\hat{T}_{g_i}\hat{T}_{g_j}e_k = \hat{T}_{g_i g_j}e_k = \sum_f D_{fk}(g_i g_j)\,e_f \tag{3.5}$$

Comparison of the last two results will show that (3.3) is, in fact, valid. We shall now say that the matrices $D(g_i)$ form a representation of order n of the group G. The space R_n is the representation space, and the basis of this space is the basis of the representation. By operating with $\hat{T}_{g_i}$ on an arbitrary vector x of the space R_n we obtain

$$\hat{T}_{g_i}x = \sum_k x_k \hat{T}_{g_i}e_k = \sum_{k,r} x_k D_{rk}(g_i)\,e_r = \sum_r x'_r e_r \tag{3.6}$$

where $x'_r = \sum_k D_{rk}(g_i)\,x_k$. Let us now consider the change in the representation matrix which occurs when a new basis e'_i is taken in the space R_n where the new basis is related to e_k by the linear transformation

$$e'_i = \sum_k V_{ki}e_k, \quad e_i = \sum_k \{V^{-1}\}_{ki}\,e'_k \tag{3.7}$$

To this end, let us apply the operator $\hat{T}_{g_i}$ to e'_j. Using (3.7) we have

$$\hat{T}_{g_i}e'_j = \sum_k V_{kj}\hat{T}_{g_i}e_k = \sum_{k,s} V_{kj}D_{sk}(g_i)\,e_s$$

$$= \sum_{k,s,r} V_{kj}D_{sk}\{V^{-1}\}_{rs}\,e'_r = \sum_r \{V^{-1}DV\}_{rj}\,e'_r \tag{3.8}$$

Thus, the representation matrices undergo a similarity

transformation when we transform to the new basis. The representation by the matrices $V^{-1}DV$ is equivalent to the representation by the matrices D.

If the representation matrices are all unitary, the representation is said to be unitary.

If the matrix group $D(g_i)$ is isomorphic to the group G, the matrices are said to give a faithful representation of the group G.

3.2 Examples of representations

Among the representations of a group there is always the trivial representation in which each element of the group is associated with the unit matrix. If the group elements are linear transformations, the matrices of these transformations themselves form a representation which is isomorphic to the group. These two representations correspond to the trivial invariant sub-groups which were mentioned in Chapter 1.

To illustrate other representations of a group, consider the derivation of one of the representations of the group C of matrices of linear transformations of n variables $x_1, x_2, \ldots, x_n$:

$$x'_i = \sum_k C_{ik} x_k \tag{3.9}$$

Consider the quadratic form

$$\sum_{i,k} a_{ik} x_i x_k, \quad a_{ik} = a_{ki} \tag{3.10}$$

Transformation of the variables $x_1, x_2, \ldots, x_n$ induces a transformation of the coefficients of this form. In point of fact, if we substitute

$$x_j = \sum_s \{C^{-1}\}_{js} x'_s \tag{3.11}$$

we obtain an expression for the quadratic form (3.10) in new (primed) variables

$$\sum_{i,k,j,l} a_{ik} \{C^{-1}\}_{ij} x'_j \{C^{-1}\}_{kl} x'_l = \sum_{j,l} a'_{jl} x'_j x'_l \tag{3.12}$$

where

$$a'_{jl} = \sum_{i,k} \{C^{-1}\}_{ij} a_{ik} \{C^{-1}\}_{kl} \tag{3.13}$$

If we write $\| a_{ik} \| = A$, we can write down the transformation rule for the coefficients a_{ik} in the matrix form

$$A' = C^{-1'} AC^{-1} \tag{3.14}$$

where $C^{-1'}$ is the transpose of C^{-1}. Let us now apply successively the transformations C_1 and C_2 to the variables x_1, x_2, ..., x_n. This yields

$$A'' = C_2^{-1'} A' C_2^{-1} = C_2^{-1'} C_1^{-1'} A C_1^{-1} C_2^{-1}$$

or

$$A'' = (C_2 C_1)^{-1'} A (C_2 C_1)^{-1} \tag{3.15}$$

We then see that the application of the transformation C_1 and then of the transformation C_2 is equivalent to the application of the transformation C_2C_1. We may thus conclude that the transformations of the coefficients of the quadratic form given by (3.13) form a representation of the group C.

3.3 Representation of the symmetry group of the Schroedinger equation, realized on its eigenfunctions

Since our main aim is to review the applications of group-theoretical methods to physical problems, it will be useful to indicate the importance of group representations to these applications. As an example, consider a quantum-mechanical system described by the Schroedinger equation

$$\left[-\frac{\hbar^2}{2m}\Delta + V(\boldsymbol{r})\right]\psi(\boldsymbol{r}) = E\psi(\boldsymbol{r}) \tag{3.16}$$

We shall assume that the symmetry group for this system consists of orthogonal transformations u_s defined by

$$\boldsymbol{r}' = u_s \boldsymbol{r} \tag{3.17}$$

We know from Chapter 1 that the substitution

$$\boldsymbol{r} = u_s^{-1} \boldsymbol{r}' \tag{3.18}$$

should conserve the form of (3.16). Since the Laplace operator is invariant under any orthogonal transformations of the coordinates, this substitution yields

$$\left[-\frac{\hbar^2}{2m}\Delta_{r'}+V\left(u_s^{-1}r\right)\right]\psi\left(u_s^{-1}r\right)=E\psi\left(u_s^{-1}r\right) \tag{3.19}$$

Moreover, since the Schroedinger equation is invariant under the transformations u_s, we must have

$$V\left(u_s^{-1}r\right)=V(r) \tag{3.20}$$

and therefore the transformed wave function

$$\psi'(r')=\hat{T}_{u_s}\psi(r)\equiv\psi\left(u_s^{-1}r\right) \tag{3.21}$$

is also an eigenfunction of the Schroedinger equation (3.9) with the same eigenvalue E. Let $\psi_1(r), \ldots, \psi_k(r)$ be a complete set of orthonormal eigenfunctions of this equation, corresponding to the eigenvalue E. We shall show that these functions form the basis of a group representation. In fact, each of the transformed functions $\hat{T}_{u_s}\psi_i(r)$ can be written in the form

$$\hat{T}_{u_s}\psi_i(r)=\psi_i\left(u_s^{-1}r\right)=\sum_{j=1}^{k}D_{ji}(u_s)\,\psi_j(r) \tag{3.22}$$

The functions $\hat{T}_{u_s}\psi_i(r)$ $(i=1, 2, \ldots, k)$ must also be orthonormal, since a change of the variable through the orthogonal transformation (3.18) conserves the orthonomalization condition:

$$\int\bar{\psi}_i\left(u_s^{-1}r\right)\psi_j\left(u_s^{-1}r\right)d\tau=\int\bar{\psi}_i(r)\,\psi_j(r)\,d\tau=\delta_{ij} \tag{3.23}$$

It follows that the matrices $\|D_{ij}(u_s)\|$ should be unitary, and, hence, to each transformation u_s from the symmetry group of the Schroedinger equation one can assign a unitary matrix of order k. We shall show that these matrices form a group representation. Let u_s and u_t be transformations in the group. Their successive application yields

$$\begin{aligned}\hat{T}_{u_s}\hat{T}_{u_t}\psi_i(r)&=\hat{T}_{u_s}\psi_i\left(u_t^{-1}r\right)=\psi_i\left(u_t^{-1}u_s^{-1}r\right)\\&=\psi_i\left((u_su_t)^{-1}r\right)=\sum_{l=1}^{k}D_{li}(u_su_t)\,\psi_l(r)\end{aligned} \tag{3.24}$$

On the other hand,

$$\hat{T}_{u_s}\hat{T}_{u_t}\psi_l(\boldsymbol{r})=\hat{T}_{u_s}\sum_{j=1}^{k}D_{jl}(u_t)\psi_j(\boldsymbol{r})=\sum_{j=1}^{k}D_{jl}(u_t)\sum_{l=1}^{k}D_{lj}(u_s)\psi_l(\boldsymbol{r})$$

$$=\sum_{l=1}^{k}\{D(u_s)D(u_t)\}_{li}\psi_l(\boldsymbol{r}) \tag{3.25}$$

If we compare (3.24) with (3.25), we find that

$$D(u_s u_t)=D(u_s)D(u_t) \tag{3.26}$$

which was to be proved.

The importance of group representations in this problem lies in that with each energy eigenvalue we can associate a representation and establish the possible types of symmetry of the wave functions without explicitly solving the Schroedinger equation.

Let us now proceed to a study of the properties of representations of finite groups.

3.4 Existence of an equivalent unitary representation

We shall show that any representation of a finite group is equivalent to a unitary representation.

Suppose that we have a representation D of the group G consisting of the m elements $g_1, g_2, \ldots, g_m$. We shall regard the representation matrices $D(g_i)$ as the transformation matrices in an n-dimensional space R_n. Let $\boldsymbol{x}(x_1, x_2, \ldots, x_n)$ and $\boldsymbol{y}(y_1, y_2, \ldots, y_n)$ be vectors in this space. The scalar product of the vectors will be defined as usual:

$$(\boldsymbol{x}, \boldsymbol{y})=x_1\bar{y}_1+x_2\bar{y}_2+\ldots+x_n\bar{y}_n \tag{3.27}$$

The transformation $D(g_i)$ transforms the vector $\boldsymbol{x}$ into the vector $\boldsymbol{x}^{(i)}$:

$$\boldsymbol{x}^{(i)}=D(g_i)\boldsymbol{x},\quad x_\alpha^{(i)}=\sum_{\beta=1}^{n}D_{\alpha\beta}(g_i)x_\beta \tag{3.28}$$

while the vector $\boldsymbol{y}$ is transformed into $\boldsymbol{y}^{(i)}$:

$$\boldsymbol{y}^{(i)}=D(g_i)\boldsymbol{y} \tag{3.29}$$

Let us suppose that the transformation $D(g_l)$ is not unitary and, consequently, it does not conserve the scalar product $(\boldsymbol{x}, \boldsymbol{y})$. We shall show that it is possible to choose a new basis in the space R_n such that the transformation matrices for the vector components in this space will be unitary. To prove this, let us take the average of the scalar product (3.27) over the group, i.e. let us construct the expression

$$\sum_{l=1}^{m} (D(g_l)\boldsymbol{x},\ D(g_l)\boldsymbol{y}) = \sum_{l=1}^{m} (\boldsymbol{x}^{(l)},\ \boldsymbol{y}^{(l)}) \tag{3.30}$$

We shall show that (3.30) can be written in the form

$$\sum_{l=1}^{m} (\boldsymbol{x}^{(l)},\ \boldsymbol{y}^{(l)}) = (L\boldsymbol{x},\ L\boldsymbol{y}) \tag{3.31}$$

where L is a linear transformation. To do this, let us write (3.30) in the form

$$\sum_{l=1}^{m} (D(g_l)\boldsymbol{x},\ D(g_l)\boldsymbol{y}) = \left(\sum_{l=1}^{m} D^+(g_l)D(g_l)\boldsymbol{x},\ \boldsymbol{y}\right) \tag{3.32}$$

The matrix $\sum\limits_{l=1}^{m} D^+(g_l)D(g_l)$ is Hermitian and can therefore be reduced to a diagonal form through a unitary transformation V. We thus find that

$$d = V^{-1}\sum_{l=1}^{m} D^+(g_l)D(g_l)V \tag{3.33}$$

and hence

$$\sum_{l=1}^{m} D^+(g_l)D(g_l) = V d V^{-1} \tag{3.34}$$

where d is a diagonal matrix.

If we substitute $\tilde{D}(g_l) = V^{-1}D(g_l)V$, we can write

$$d = \sum_{l=1}^{m} V^{-1}D^+(g_l)VV^{-1}D(g_l)V = \sum_{l=1}^{m} \tilde{D}^+(g_l)\tilde{D}(g_l) \tag{3.35}$$

and hence the diagonal elements of the matrix d are given by

$$d_{\alpha\alpha}=\sum_{i=1}^{m}\sum_{\beta=1}^{n}\tilde{D}^{*}_{\alpha\beta}(g_i)\,\tilde{D}_{\beta\alpha}(g_i)=\sum_{i=1}^{m}\sum_{\beta=1}^{n}|\tilde{D}_{\beta\alpha}(g_i)|^2>0 \tag{3.36}$$

Let us determine the diagonal matrix $d^{1/2}$ whose elements are $\{d^{1/2}\}_{\alpha\alpha}=\sqrt{d_{\alpha\alpha}}$. It is clear that $d^{1/2}d^{1/2}=d$, and if we use the self-adjoint property of the matrix $d^{1/2}$, we have

$$\sum_{i=1}^{m}(x^{(i)},\ y^{(i)})=(V\,dV^{-1}x,\ y)$$
$$=(d^{1/2}d^{1/2}V^{-1}x,\ V^{-1}y)=(d^{1/2}V^{-1}x,\ d^{1/2}V^{-1}y) \tag{3.37}$$

We thus arrive at the equation given by (3.31), where L is given by

$$L=d^{1/2}V^{-1} \tag{3.38}$$

We can now show that the representation of G given by the matrices LDL^{-1} is unitary. To begin with we shall show that for an arbitrary element g_k of G we have

$$(LD(g_k)x,\ LD(g_k)y)=(Lx,\ Ly) \tag{3.39}$$

In fact, according to (3.30) and (3.31),

$$(LD(g_k)x,\ LD(g_k)y)=\sum_{i=1}^{m}(D(g_i)D(g_k)x,\ D(g_i)D(g_k)y)$$
$$=\sum_{i=1}^{m}(D(g_ig_k)x,\ D(g_ig_k)y) \tag{3.40}$$

However, we know that when the element g_i runs over the entire group, the element g_ig_k will also do so. We can therefore finally write

$$(LD(g_k)x,\ LD(g_k)y)=\sum_{l=1}^{m}(D(g_l)x,\ D(g_l)y)=(Lx,\ Ly) \tag{3.41}$$

If we now substitute $x'=L^{-1}x$ and $y'=L^{-1}y$ we can rewrite (3.39) in the form

$$(LD(g_k)L^{-1}x',\ LD(g_k)L^{-1}y')=(x',\ y') \tag{3.42}$$

and hence it follows that the matrices $LD(g_k)L^{-1}(g_k\in G)$ are, in fact, unitary.

3.5 Reducible and irreducible representations of a group

Suppose that a representation D of the group G is given in a space R_n. If in the space R_n there is a sub-space R_k $(k < n)$ which is invariant under all the transformations D, i.e. if for $\boldsymbol{x} \in R_k$ we have $D\boldsymbol{x} \in R_k$, the representation is reducible. Let us take the first k unit vectors in the space R_n as the unit vectors of the sub-space R_k. The representation matrix must then have the following form:

$$\left|\begin{array}{cccc|ccc} D_{11} & D_{12} & \ldots & D_{1k} & D_{1\,k+1} & \ldots & D_{1n} \\ D_{21} & D_{22} & \ldots & D_{2k} & D_{2\,k+1} & \ldots & D_{2n} \\ \cdot & \cdot & \cdot & \cdot & \cdot & \cdot & \cdot \\ D_{k1} & D_{k2} & \ldots & D_{kk} & D_{k\,k+1} & \ldots & D_{kn} \\ \hline 0 & 0 & \ldots & 0 & D_{k+1\,k+1} & \ldots & D_{k+1\,n} \\ \cdot & \cdot & \cdot & \cdot & \cdot & \cdot & \cdot \\ 0 & 0 & \ldots & 0 & D_{n\,k+1} & \ldots & D_{nn} \end{array}\right|$$

If, on the other hand, we cannot define an invariant sub-space in R_n, the representation is irreducible.

We shall show that if a reducible representation D is unitary, the orthogonal complement of the sub-space R_k, which we shall denote by R_{n-k}, is also invariant under the transformations D. In point of fact, let $\boldsymbol{x} \in R_k$, $\boldsymbol{y} \in R_{n-k}$, in which case $(\boldsymbol{x}, \boldsymbol{y}) = 0$. Since the sub-space R_k is invariant, we have

$$(D(g)\boldsymbol{x}, \boldsymbol{y}) = 0 \tag{3.43}$$

But

$$(D(g)\boldsymbol{x}, \boldsymbol{y}) = (\boldsymbol{x}, D^{+}(g)\boldsymbol{y}) = (\boldsymbol{x}, D^{-1}(g)\boldsymbol{y}) = (\boldsymbol{x}, D(g^{-1})\boldsymbol{y}) = 0 \tag{3.44}$$

and, hence,

$$D(g^{-1})\boldsymbol{y} \in R_{n-k} \tag{3.45}$$

When g runs over the entire group, the inverse element g^{-1} will also do so. Therefore, (3.45) is satisfied for all matrices of the representation in question, and the invariance of R_{n-k} is proved. If we now take the unit vectors of the sub-space R_k as the first k unit vectors, and the remaining $n-k$ unit vectors are taken as the unit vectors of the sub-space R_{n-k}, the representation matrix will have the following quasi-diagonal form:

$$\left|\begin{array}{cccc|cccc} D_{11} & D_{12} & \ldots & D_{1k} & 0 & \ldots & & 0 \\ D_{21} & D_{22} & \ldots & D_{2k} & 0 & \ldots & & 0 \\ \cdot & \cdot & \cdot & \cdot & \cdot & \cdot & \cdot & \cdot \\ D_{k1} & D_{k2} & \ldots & D_{kk} & 0 & \ldots & & 0 \\ \hline 0 & 0 & \ldots & 0 & D_{k+1\,k+1} & \ldots & & D_{k+1\,n} \\ \cdot & \cdot & \cdot & \cdot & \cdot & \cdot & \cdot & \cdot \\ 0 & 0 & \ldots & 0 & D_{n\,k+1} & \ldots & & D_{nn} \end{array}\right|$$

If the space R can be resolved into invariant sub-spaces, in each of which an irreducible representation is realized, the representation D is fully reducible. With a suitable choice of unit vectors, the matrix of this representation has the following quasi-diagonal form:

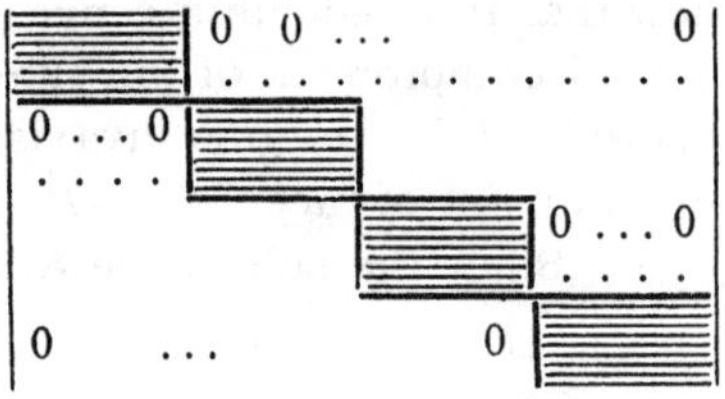

It follows from the foregoing discussion that:

1. a unitary representation of a group is always either irreducible or fully reducible;

2. any representation of a finite group is either irreducible or fully reducible (since it is equivalent to a unitary representation).

If the representation D is reducible, its matrices can be reduced to a quasi-diagonal form by going over to the new system of unit vectors, as we have seen. We note that, in this case, the representation matrices undergo the similarity transformation

$$D \to V^{-1} D V$$

where V is the matrix relating the unit vectors of the old and new bases (see Equation (3.7)). The condition that the representation is reducible can therefore be formulated as follows. A representation D is reducible if there exists a non-singular matrix V, such that the matrices $V^{-1}DV$ are quasi-diagonal.

3.6 Schur's first lemma

We shall now prove an important theorem known as Schur's first lemma:

A matrix which commutes with all the matrices of an irreducible representation is a multiple of the unit matrix.

Proof. Let $D(g)$ be the matrices of an irreducible representation of order n of the group G, $g \in G$. We shall suppose that the matrix M commutes with all the matrices $D(g)$:

$$MD(g) = D(g)M \tag{3.46}$$

Let R_n represent the space in which the representation $D(g)$ is realized. In this space there should be at least one eigenvector of the matrix M. Let us denote it by $\boldsymbol{x}$. We then have

$$M\boldsymbol{x} = \lambda \boldsymbol{x} \tag{3.47}$$

If we apply the transformation with the representation matrix $D(g)$ to the vector $\boldsymbol{x}$ we have

$$D(g)\boldsymbol{x} = \boldsymbol{x}_g \tag{3.48}$$

where the resulting vector $\boldsymbol{x}_g$ is also an eigenvector of M with the same eigenvalue λ. In fact, in view of (3.46), we have

$$M\boldsymbol{x}_g = MD(g)\boldsymbol{x} = D(g)M\boldsymbol{x} = \lambda D(g)\boldsymbol{x} = \lambda \boldsymbol{x}_g \tag{3.49}$$

Hence, it follows that the space of eigenvectors of the matrix M corresponding to the same eigenvalue is invariant under the transformations $D(g)$. But since, by hypothesis, the representation $D(g)$ is irreducible, it follows that this subspace should coincide with the entire space R_n, and the matrix M multiplying any vector of the space R_n by the number λ should be of the form

$$M = \begin{pmatrix} \lambda & 0 & 0 \dots 0 \\ 0 & \lambda & 0 \dots 0 \\ \cdot & \cdot & \cdot \cdot \cdot \cdot \\ 0 & 0 & 0 \dots \lambda \end{pmatrix}$$

This proves the theorem.

If a representation is fully reducible, i.e. its matrices have the quasi-diagonal form, one can always find a matrix which is not a multiple of the unit matrix and which commutes with all the matrices of this representation. It is readily verified that this matrix can be taken to be the diagonal matrix in which diagonal elements corresponding to different blocks of the representation matrix are not equal to one another.

Hence, it may be concluded that if the only matrix which commutes with all the matrices of a representation of a group is the matrix which is equal to a multiple of the unit matrix, then this representation is irreducible.

3.7 Schur's second lemma

Let $D^{(1)}(g)$ and $D^{(2)}(g)$ be the matrices of two irreducible non-equivalent representations of a group G of order n_1 and n_2, respectively. Then any rectangular matrix M with n_1 columns and n_2 rows which satisfies the equation

$$MD^{(1)}(g) = D^{(2)}(g)\,M \tag{3.50}$$

for all $g \in G$ is a null matrix.

Proof. Let us take the Hermitian conjugate of both sides of (3.50). This yields

$$D^{(1)+}(g)\,M^+ = M^+ D^{(2)+}(g) \tag{3.51}$$

If the representations $D^{(1)}$ and $D^{(2)}$ are unitary, then

$$D^{(1)^{-1}}(g)\, M^{+} = M^{+} D^{(2)^{-1}}(g) \tag{3.52}$$

or

$$D^{(1)}(g^{-1})\, M^{+} = M^{+} D^{(2)}(g^{-1}) \tag{3.53}$$

If the element g runs over the entire group, then g^{-1} will also do so. The last equation can therefore be written in the form

$$D^{(1)}(g)\, M^{+} = M^{+} D^{(2)}(g) \tag{3.54}$$

Let us multiply both sides of this equation on the left by the matrix M:

$$MD^{(1)}(g)\, M^{+} = MM^{+} D^{(2)}(g)$$

Using (3.50) we have

$$D^{(2)}(g)\, MM^{+} = MM^{+} D^{(2)}(g). \tag{3.55}$$

and, hence, according to Schur's first lemma, we conclude that the MM^{+} must be a multiple of a unit matrix:

$$MM^{+} = \lambda E_{n_2} \tag{3.56}$$

where E_{n_2} is a unit matrix of order n_2.

Let us now consider three possible special cases.

1. $n_1 = n_2$. In this case, the matrix M must be singular, i.e. $\det M = 0$. In fact, if this were not so, then (3.50) would yield the condition for the equivalence of representations:

$$D^{(1)}(g) = M^{-1} D^{(2)}(g)\, M \tag{3.57}$$

If we evaluate the determinants of both sides of (3.56), we obtain

$$\det M \det M^{+} = \lambda^{n_2} = 0 \tag{3.58}$$

and hence $\lambda = 0$. On the other hand, from (3.56) we have

$$\lambda = \sum_j M_{ij} \overline{M}_{ij} = \sum_j | M_{ij} |^2 \tag{3.59}$$

and, consequently, λ can vanish only if the matrix elements of M_{ij} are zero.

2. $n_2 > n_1$. Let us augment the matrix M so that it becomes a square matrix with $n_2 - n_1$ zero columns, and let

us apply the same operation to M^+ so that it has the same number of zero rows. If we denote the two new matrices by $\tilde{M}$ and $\tilde{M}^+$, then it is clear that (3.56) will be satisfied for them, i.e.

$$\tilde{M}\tilde{M}^+ = \lambda E_{n_2} \tag{3.60}$$

By definition,

$$\det \tilde{M} = \det \tilde{M}^+ = 0$$

and therefore, if we repeat the discussion given for the first case, we again find that

$$M_{ik} = 0 \tag{3.61}$$

3. $n_2 < n_1$. This case reduces to case (2), and we leave it to the reader to discuss it.

In proving Schur's lemmas we used the fact that the representations $D^{(1)}$ and $D^{(2)}$ were unitary. We shall now show that this limitation is unimportant. We know that any representation of a finite group is equivalent to a unitary representation. Suppose, for example, that $D^{(1)}$ and $D^{(2)}$ are non-unitary representations. It is always possible to find non-singular matrices V and W, such that the representations

$$\tilde{D}^{(1)} = V^{-1}D^{(1)}V, \quad \tilde{D}^{(2)} = W^{-1}D^{(2)}W \tag{3.62}$$

will be unitary. The condition given by (3.50) can then be written in the form

$$MV\tilde{D}^{(1)}V^{-1} = W\tilde{D}^{(2)}W^{-1}M \tag{3.63}$$

and hence

$$(W^{-1}MV)\tilde{D}^{(1)} = \tilde{D}^{(2)}(W^{-1}MV) \tag{3.64}$$

or, on substituting $N = W^{-1}MV$,

$$N\tilde{D}^{(1)} = \tilde{D}^{(2)}N \tag{3.65}$$

The problem is thus reduced to that discussed above. The matrix N can only be a null matrix. However, if this is so, the matrix $M = WNV^{-1}$ will necessarily be a null matrix.

3.8 Orthogonality relations for matrix elements of irreducible representations

Schur's lemmas can be used to establish certain relations between the matrix elements of irreducible representations of a group.

Let $D^{(i)}(g)$ and $D^{(j)}(g)$ be the matrices of two irreducible non-equivalent unitary representations of a group G consisting of m elements. Let n_i and n_j be the orders of these representations. We shall show that the following relationships exist between the elements of the matrices $D^{(i)}$ and $D^{(j)}$:

$$\sum_{g \in G} D^{(i)}_{\mu\nu}(g)\, \bar{D}^{(j)}_{\alpha\beta}(g) = 0 \tag{3.66}$$

$$\sum_{g \in G} D^{(i)}_{\mu\nu}(g) \bar{D}^{(i)}_{\alpha\beta}(g) = \frac{m}{n_i} \delta_{\mu\alpha} \delta_{\nu\beta} \tag{3.67}$$

Proof. Consider the matrix

$$M = \sum_{g \in G} D^{(i)}(g)\, X\, D^{(j)}(g^{-1})$$

where X is an arbitrary matrix with n_i rows and n_j columns. We shall prove that

$$D^{(i)} M = M D^{(j)} \tag{3.68}$$

In fact,

$$D^{(i)}(g')\, M = D^{(i)}(g') \sum_{g \in G} D^{(i)}(g)\, X D^{(j)}(g^{-1})$$

$$= \sum_{g \in G} D^{(i)}(g')\, D^{(i)}(g)\, X D^{(j)}(g^{-1})\, D^{(j)}(g'^{-1})\, D^{(j)}(g')$$

$$= \sum_{g \in G} D^{(i)}(g'g)\; X D^{(j)}\left((g'g)^{-1}\right) D^{(j)}(g')$$

$$= \sum_{g'' \in G} D^{(i)}(g'')\, X D^{(j)}(g''^{-1})\, D^{(j)}(g') = M D^{(j)}(g')$$

and hence it follows from Schur's second lemma that M is a null matrix, i.e.

$$M_{\mu\alpha}=\sum_{g\in G}\sum_{s,\,k}D_{\mu s}^{(i)}(g)\,X_{sk}D_{k\alpha}^{(j)}(g^{-1})=0 \tag{3.69}$$

Since the matrix X is arbitrary, we can set $X_{sk}=1$ if $s=\nu$ and $k=\beta$, and $X_{sk}=0$ for other values of s and k. We then have

$$\sum_{g\in G}D_{\mu\nu}^{(i)}(g)\,D_{\beta\alpha}^{(j)}(g^{-1})=0 \tag{3.70}$$

We note that, so far, we have not assumed that the representations were unitary and, therefore, (3.70) is also valid for non-unitary representations. If, on the other hand, the representations $D^{(i)}$ and $D^{(j)}$ are unitary, then (3.66) follows from (3.70).

Let us now go on to prove the second orthogonality relation.

Consider the matrix

$$N=\sum_{g\in G}D^{(i)}(g)\,XD^{(i)}(g) \tag{3.71}$$

where X is an arbitrary square matrix of order n_i. It can be shown in a similar fashion that it commutes with all the matrices of the irreducible representation $D^{(i)}$. Consequently, by Schur's first lemma the matrix N is a multiple of the unit matrix, i.e.

$$N_{\mu\alpha}=\sum_{g\in G}\sum_{s,\,k}D_{\mu s}^{(i)}(g)\,X_{sk}D_{k\alpha}^{(i)}(g^{-1})=\lambda\delta_{\mu\alpha} \tag{3.72}$$

Let us now select a matrix X in which the only non-zero element $X_{\nu\beta}$ is equal to unity. The corresponding constant λ will be denoted by $\lambda_{\nu\beta}$. We obtain

$$\sum_{g\in G}D_{\mu\nu}^{(i)}(g)\,D_{\beta\alpha}^{(i)}(g^{-1})=\lambda_{\nu\beta}\delta_{\mu\alpha} \tag{3.73}$$

To determine $\lambda_{\nu\beta}$ let us substitute $\mu=\alpha$ into this equation and take the sum of both of its sides over μ between 1 and n_i. This yields

$$\sum_{g\in G}\sum_{\mu}D_{\mu\nu}^{(i)}(g)\,D_{\beta\mu}^{(i)}(g^{-1})=\lambda_{\nu\beta}n_i \tag{3.74}$$

or

$$\sum_{g \in G} D^{(i)}_{\beta\nu}(E) = \delta_{\nu\beta} m = \lambda_{\nu\beta} n_i \tag{3.75}$$

Hence we find that

$$\lambda_{\nu\beta} = \delta_{\nu\beta} \frac{m}{n_i} \tag{3.76}$$

and, therefore,

$$\sum_{g \in G} D^{(i)}_{\mu\nu}(g) D^{(i)}_{\beta\alpha}(g^{-1}) = \delta_{\mu\alpha}\delta_{\nu\beta} \frac{m}{n_i} \tag{3.77}$$

If the representation $D^{(i)}(g)$ is unitary, then (3.67) follows from (3.77). The two orthogonality relations (3.66) and (3.67) which we have proved above can be combined into the single form

$$\sum_{g \in G} D^{(i)}_{\mu\nu}(g) \bar{D}^{(j)}_{\alpha\beta}(g) = \frac{m}{n_i} \delta_{ij}\delta_{\mu\alpha}\delta_{\nu\beta} \tag{3.78}$$

This can be interpreted as the orthogonality and normalization condition for a set of vectors in m-dimensional space. Each of these vectors is characterized by three subscripts i, μ, ν, and their components are equal to the elements of the matrices of non-equivalent irreducible representations. For example, the vector $D^{(i)}_{\mu\nu}$ has the components $D^{(i)}_{\mu\nu}(g_1)$, $D^{(i)}_{\mu\nu}(g_2)$, . . . , $D^{(i)}_{\mu\nu}(g_m)$. The number of such vectors which correspond to one reducible representation, say $D^{(i)}$, is equal to n_i^2, and therefore the total number of orthonormal vectors in this system is

$$\sum_i n_i^2 \tag{3.79}$$

where the sum is evaluated only over the non-equivalent irreducible representations. Because of orthogonality, all these vectors should be linearly independent. Since the number of linearly independent vectors cannot exceed the dimensionality of the vector space, it follows that

$$\sum_i n_i^2 \leqslant m \tag{3.80}$$

We have thus deduced the important result that the number

of different irreducible representations of a finite group is finite. It will be shown in Section 3.10 that the equation

$$\sum_i n_i^2 = m \tag{3.81}$$

is always valid.

3.9 Characters of representations

The character of a representation $D(g)$ is defined as the following function of the group elements:

$$\chi(g) = \sum_i D_{ii}(g) = \mathrm{Sp}\, D(g) \tag{3.82}$$

Let us consider some of the properties of the characters of representations.

a. Equivalent representations have identical characters, since the trace of a matrix is invariant under the similarity transformation and, consequently, $\mathrm{Sp}\, V^{-1} D(g) V = \mathrm{Sp}\, D(g)$.

b. The characters of the representation matrices corresponding to the elements of a given class are identical.

c. The characters of the irreducible representations have the orthogonality property:

$$\sum_{g \in G} \chi^{(i)}(g) \bar{\chi}^{(j)}(g) = m\delta_{ij} \tag{3.83}$$

where $\chi^{(i)}(g)$ and $\chi^{(j)}(g)$ are the characters of the irreducible representations $D^{(i)}$ and $D^{(j)}$, respectively. In view of property (a), it is sufficient to prove (3.83) for unitary representations. From (3.78) we have

$$\sum_{g \in G} D^{(i)}_{\mu\mu}(g) \bar{D}^{(j)}_{\alpha\alpha}(g) = \frac{m}{n_i} \delta_{ij} \delta_{\mu\alpha} \tag{3.84}$$

and if we sum both sides of this equation over μ and α, we obtain

$$\sum_{g \in G} \chi^{(i)}(g) \bar{\chi}^{(j)}(g) = \frac{m}{n_i} n_i \delta_{ij} = m\delta_{ij} \tag{3.85}$$

which was to be proved. The function $\chi(g)$ has the same value for all the elements of a given class. Therefore, the above relationship can also be written in the form

$$\sum_s k_s \chi_s^{(i)} \bar{\chi}_s^{(j)} = m\delta_{ij} \tag{3.86}$$

where k_s is the number of elements in class C_s, and $\chi_s^{(i)}$ is the value of the character of the representation corresponding to the elements of this class.

d. The character of a reducible representation D is equal to the sum of characters of irreducible representations into which it can be decomposed. To show this, it is sufficient to recall the quasi-diagonal form of the reducible representation and also property (a). If we denote the character of the reducible representation by $\chi(g)$, then

$$\chi(g) = \sum_j r_j \chi^{(j)}(g) \tag{3.87}$$

where the number r_j shows how many times the irreducible representation $D^{(j)}$ enters the decomposition of the reducible representation D. From the orthogonality relation we can readily establish the formula

$$r_j = \frac{1}{m} \sum_{g \in G} \bar{\chi}(g) \chi^{(j)}(g) \tag{3.88}$$

which is important for applications. Hence, it follows inter alia that the decomposition of a reducible representation into irreducible parts can be carried out uniquely.

The decomposition of a reducible representation D into irreducible representations can be symbolically written in the form

$$D = \sum_j^{\oplus} r_j D^{(j)}$$

where the symbol $\oplus$ reminds us that the expression on the right-hand side of the equation is not the sum of matrices in the usual sense.

3.10 The regular representation

Let us take an arbitrary element g_s of a given group G and perform the operation of translation over the group, i.e. let us multiply each of the group elements on the left by g_s. In accordance with Section 2.1 we then find that, since $g_s \neq E$, none of the group elements remains in its place. If, on the other hand, $g_s = E$, there is no translation.

The translation corresponding to an element g_s can be formally written in the form

$$g_s g_l = \sum_t R_{tl}(g_s)\, g_t \tag{3.89}$$

where $\| R_{lj}(g_s) \|$ is a matrix of order m. It is clear that in each column of this matrix there is only one non-zero element (equal to unity). If $g_s g_l = g_j$, then $R_{jl}(g_s) = 1$, and $R_{tl}(g_s) = 0$ for $t \neq j$. The matrices $R(g_s)$ constructed in this way provide the regular representation of order m of the group G.

It follows from the above definition that the characters of the regular representation are given by

$$\left.\begin{aligned} \chi^{(R)}(g_s) &= m \qquad \text{if} \quad g_s = E \\ \chi^{(R)}(g_s) &= 0 \qquad \text{if} \quad g_s \neq E \end{aligned}\right\} \tag{3.90}$$

Let us decompose the regular representation into irreducible parts, i.e. establish how many times it contains each irreducible representation $D^{(j)}$. To do this, let us use (3.88). We have

$$r_j = \frac{1}{m} \sum_g \bar{\chi}^{(R)}(g)\, \chi^{(j)}(g) \tag{3.91}$$

or, according to (3.90),

$$r_j = \frac{1}{m}\, m \chi^{(j)}(E) = \chi^{(j)}(E) = n_j \tag{3.92}$$

We thus see that the number of times each irreducible

representation occurs in the regular representation is equal to the order of this irreducible representation.

We can use this theorem to express the order of a regular representation in terms of the order of the irreducible representations into which it can be decomposed. We thus have

$$\sum_j n_j^2 = m \tag{3.93}$$

We recall that the expression on the left of this equation gives the number of orthogonal vectors $D^{(j)}_{\alpha\beta}$. This number is equal to the dimensionality of the vector space and, therefore, the vectors $D^{(j)}_{\alpha\beta}$ form a complete set in this space. We shall use this result in the next section to establish a theorem on the number of irreducible representations of a finite group.

3.11 The number of irreducible representations

The characters $\chi(g_1)$, $\chi(g_2)$, ..., $\chi(g_m)$ of a representation can also be regarded as the components of a vector in an m-dimensional space R_m. It follows from (3.83) that the characters of irreducible representations then form a set of orthogonal vectors. Since for the elements of a given class the characters are the same, it follows that all such vectors belong to a sub-space $R_\varkappa$ of R_m. The space $R_\varkappa$ is characterized by the fact that the components of vectors corresponding to elements of a given class are equal. The components of an arbitrary vector $\boldsymbol{F}(F(g_1), F(g_2), \ldots)$ of the sub-space $R_\varkappa$ have the property

$$F(g') = F(g^{-1}g'g) \tag{3.94}$$

for any g' and g belonging to the group G. Since the number of different components of the vector $\boldsymbol{F}$ cannot exceed the number of classes in the group G, the maximum number $\varkappa$ of linearly independent vectors in $R_\varkappa$ is equal to the number of classes in the group. We shall show that an arbitrary vector F in the sub-space $R_\varkappa$ can be resolved in terms of the vectors $\chi^{(j)}$ corresponding to irreducible representations of the group G.

Since the vector F belongs to the space R_m, it can be resolved in terms of the complete set of vectors $D^{(j)}_{\alpha\beta}$, and for the components of this vector we have

$$F(g') = \sum_{j,\alpha,\beta} C^{(j)}_{\alpha\beta} D^{(j)}_{\alpha\beta}(g') \tag{3.95}$$

Using (3.94) we can write

$$F(g') = F(g^{-1}g'g) = \sum_{j,\alpha,\beta} C^{(j)}_{\alpha\beta} D^{(j)}_{\alpha\beta}(g^{-1}g'g) \tag{3.96}$$

and if we take the average of this equation with respect to g we obtain

$$\begin{aligned} F(g') &= \frac{1}{m}\sum_{g}\sum_{j,\alpha,\beta} C^{(j)}_{\alpha\beta} D^{(j)}_{\alpha\beta}(g^{-1}g'g) \\ &= \frac{1}{m}\sum_{g}\sum_{j,\alpha,\beta} C^{(j)}_{\alpha\beta} \sum_{\gamma,\delta} D^{(j)}_{\alpha\gamma}(g^{-1}) D^{(j)}_{\gamma\delta}(g') D^{(j)}_{\delta\beta}(g) \\ &= \frac{1}{m}\sum_{g}\sum_{j,\alpha,\beta,\gamma,\delta} C^{(j)}_{\alpha\beta} \overline{D}^{(j)}_{\gamma\alpha}(g) D^{(j)}_{\delta\beta}(g) D^{(j)}_{\gamma\delta}(g') \end{aligned} \tag{3.97}$$

Substituting the orthogonality relations (3.67), we have

$$\begin{aligned} F(g') &= \frac{1}{m}\sum_{j,\alpha,\beta,\gamma,\delta} C^{(j)}_{\alpha\beta} \frac{m}{n_j} \delta_{\gamma\delta}\delta_{\alpha\beta} D^{(j)}_{\gamma\delta}(g') \\ &= \frac{1}{m}\sum_{j,\alpha,\delta} C^{(j)}_{\alpha\alpha} \frac{m}{n_j} D^{(j)}_{\delta\delta}(g') = \sum_{j} B_j \chi^{(j)}(g') \end{aligned} \tag{3.98}$$

where

$$B_j = \frac{1}{n_j}\sum_{\alpha} C^{(j)}_{\alpha\alpha}$$

We thus see that an arbitrary vector $F \in R_\varkappa$ can be resolved in terms of the vectors $\chi^{(j)}$. Hence it follows that the set of vectors $\chi^{(j)}$ is complete in $R_\varkappa$ and, consequently, the number of these vectors is equal to the number $\varkappa$ of classes in G.

We thus arrive at the following important theorem: the number of irreducible representations of a group is equal to the number of its classes.

3.12 Determination of the characters of the irreducible representations

We note that the characters of irreducible representations satisfy the orthogonality and normalization relations (3.83). Moreover, if we subdivide the order m of a group into $\varkappa$ squares of integers, we can determine from (3.93) the orders of the irreducible representations, which are, of course, equal to the characters of the representations of the unit group element. However, in general, these conditions are insufficient for a single-valued determination of all the characters of irreducible representations.

We shall now show that it is possible to obtain additional quadratic relations for the characters of irreducible representations, which will enable us to solve the problem. It was shown in Chapter 2 that all the possible products of the elements of two classes of a group form a set consisting of whole classes of this group. This result was written in the form (see Equation (2.5)):

$$C_i C_j = \sum_k h_{ijk} C_k \tag{3.99}$$

Consider one of the irreducible representations $D^{(p)}$ of order n_p of the group G. Let us construct the products of matrices of this representation corresponding to the products of the elements forming the set $C_i C_j$:

$$D^{(p)}(g_i)\, D^{(p)}(g_j), \quad g_i \in C_i, \; g_j \in C_j$$

We must now write down the matrices of the representation $D^{(p)}$ corresponding to the elements of the set $\sum_k h_{ijk} C_k$. It is clear that in view of (3.99) the sets of representation matrices constructed in this way should coincide. Moreover, it is clear that

$$\sum_{\substack{g_i \in C_i \\ g_j \in C_j}} D^{(p)}(g_i)\, D^{(p)}(g_j) = \sum_{g_k \in C_k} h_{ijk} D^{(p)}(g_k) \tag{3.100}$$

Writing

$$S_i^{(p)} = \sum_{g_i \in C_i} D^{(p)}(g_i) \tag{3.101}$$

in (3.100), we have

$$S_i^{(p)} S_j^{(p)} = \sum_k h_{ijk} S_k^{(p)} \tag{3.102}$$

However, the matrices $S_i^{(p)}$ are multiples of unit matrices (see Exercise 3.3):

$$S_i^{(p)} = \lambda_i^{(p)} E_{n_p} \tag{3.103}$$

If we take the trace of the matrix $S_i^{(p)}$, we have

$$\mathrm{Sp}\, S_i^{(p)} = \sum_{g_i \in C_i} \mathrm{Sp}\, D^{(p)}(g_i) = k_i \chi_i^{(p)} \tag{3.104}$$

where k_i is the number of elements in the class C_i, while, on the other hand,

$$\mathrm{Sp}\, S_i^{(p)} = n_p \lambda_i^{(p)} \tag{3.105}$$

and, therefore,

$$\lambda_i^{(p)} = \frac{k_i}{n_p} \chi_i^{(p)} \tag{3.106}$$

Substituting (3.103) into (3.102) and using (3.106), we obtain

$$k_i k_j \chi_i^{(p)} \chi_j^{(p)} = n_p \sum_l h_{ijl} k_l \chi_l^{(p)} \tag{3.107}$$

These equations must be satisfied by the characters of irreducible representations. As a rule, they need not be solved in practice because the characters of the representations of most finite groups used in applications have been calculated and tabulated.

Exercises

3. 1. Prove that any representation of a simple group, i.e. one without a normal divisor, is isomorphic to the group itself.

3. 2. Use Schur's first lemma to show that all irreducible representations of an Abelian group are of order one.

3. 3. Use Schur's first lemma to show that the sum of the irreducible representation matrices corresponding to the elements of a given class is a multiple of the unit matrix.

3. 4. Construct the matrices of the regular representation for a group of order six (see Exercise 2. 1).

3. 5. Show that, if two elements of a group are mutually inverse, then the characters of their representation are mutually complex conjugate.

3. 6. Show that the equation $\frac{1}{m}\sum_{g}\bar{\chi}(g)\chi(g) = 1$ is a sufficient condition for a representation to be irreducible.

3. 7. Show that the sum over a group of the matrix elements of any irreducible representation other than the identity representation is equal to zero.

Chapter 4

Composition of Representations and the Direct Products of Groups

Before we discuss applications, let us introduce two further concepts, namely, the concept of the composition or the direct product of group representations, and the concept of the direct product of groups. With this in view, let us first consider the direct product of matrices.

4.1 Direct product of matrices

Consider two square matrices A and B of order n and m, respectively, with elements

$$\left.\begin{array}{ll} a_{ik} & (i,\ k=1,\ 2,\ \ldots,\ n) \\ b_{\alpha\beta} & (\alpha,\ \beta=1,\ 2,\ \ldots,\ m) \end{array}\right\} \qquad (4.1)$$

The direct product of the matrices A and B is the super-matrix $A \times B$ of order n, of which the $(i,\ k)$-th element is the matrix $a_{ik}B$ of order m. As an example, let us write down the direct product of two 2×2 matrices:

$$\begin{pmatrix} a_{11} & a_{12} \\ a_{21} & a_{22} \end{pmatrix} \times \begin{pmatrix} b_{11} & b_{12} \\ b_{21} & b_{22} \end{pmatrix} = \begin{pmatrix} a_{11}B & a_{12}B \\ a_{21}B & a_{22}B \end{pmatrix} =$$

$$= \begin{pmatrix} a_{11}b_{11} & a_{11}b_{12} & a_{12}b_{11} & a_{12}b_{12} \\ a_{11}b_{21} & a_{11}b_{22} & a_{12}b_{21} & a_{12}b_{22} \\ a_{21}b_{11} & a_{21}b_{12} & a_{22}b_{11} & a_{22}b_{12} \\ a_{21}b_{21} & a_{21}b_{22} & a_{22}b_{21} & a_{22}b_{22} \end{pmatrix} \tag{4.2}$$

We see that the elements of the matrix $A \times B$ are all the possible products of the elements of A and B. It is convenient to use two indices to identify the rows and columns of the direct product of two matrices:

$$\{A \times B\}_{i\alpha,\, k\beta} = a_{ik} b_{\alpha\beta} \tag{4.3}$$

It is clear that the order of the direct product of matrices is equal to the product of the orders of the multiplicands.

It follows from the above definition that the direct product of diagonal matrices will also be a diagonal matrix, and the direct product of unit matrices is a unit matrix.

Consider now some of the properties of the direct product of two matrices.

a. If $A^{(1)}$ and $A^{(2)}$ are matrices of order n, and $B^{(1)}$ and $B^{(2)}$ are matrices of order m, then

$$(A^{(1)} \times B^{(1)})(A^{(2)} \times B^{(2)}) = A^{(1)}A^{(2)} \times B^{(1)}B^{(2)} \tag{4.4}$$

To prove this, let us write down the $(i\alpha,\, k\beta)$-th element for the left- and right-hand sides of this equation. The element of the matrix on the left is

$$\{(A^{(1)} \times B^{(1)})(A^{(2)} \times B^{(2)})\}_{i\alpha,\, k\beta}$$

$$= \sum_{l,\delta} \{A^{(1)} \times B^{(1)}\}_{i\alpha,\, l\delta} \{A^{(2)} \times B^{(2)}\}_{l\delta,\, k\beta} = \sum_{l,\delta} a^{(1)}_{il} b^{(1)}_{\alpha\delta} a^{(2)}_{lk} b^{(2)}_{\delta\beta} \tag{4.5}$$

while the element of the matrix on the right of (4.4) can be written in the form

$$\{(A^{(1)}A^{(2)}) \times (B^{(1)}B^{(2)})\}_{i\alpha,\, k\beta} = \{A^{(1)}A^{(2)}\}_{ik} \{B^{(1)}B^{(2)}\}_{\alpha\beta}$$

$$= \sum_{l} a^{(1)}_{il} a^{(2)}_{lk} \sum_{\delta} b^{(1)}_{\alpha\delta} b^{(2)}_{\delta\beta} = \sum_{l,\delta} a^{(1)}_{il} a^{(2)}_{lk} b^{(1)}_{\alpha\delta} b^{(2)}_{\delta\beta} \tag{4.6}$$

It follows that it corresponds to the element of the matrix on the left of (4.4) and, therefore, the formula given by (4.4) is valid.

b. If the matrices A and B are unitary, the matrix $A \times B$ is also unitary.

To prove this, we note that it follows from property (a) that

$$(A \times B)^{-1} = A^{-1} \times B^{-1} \tag{4.7}$$

On the other hand, it is obvious that

$$(A \times B)^{+} = A^{+} \times B^{+} \tag{4.8}$$

Since A and B are unitary, it follows that $A^{+} = A^{-1}$, $B^{+} = B^{-1}$, and

$$(A \times B)^{+} = A^{+} \times B^{+} = A^{-1} \times B^{-1} = (A \times B)^{-1} \tag{4.9}$$

which was to be proved.

We have so far introduced the concept of the direct product for square matrices. It is occasionally useful to define the direct product for rectangular matrices, which is introduced in the same way. It is clear that it is possible to have the direct product of more than two matrices.

4.2 Composition of group representations

We are now in a position to introduce the concept of the composition, or the direct product, of group representations.

Suppose that we have two representations D and D' (not necessarily irreducible) of a group G. We shall consider the matrices of these representations as the transformation matrices for l_1- and l_2-dimensional spaces R_{l_1} and R_{l_2}. For the basis vectors $\boldsymbol{u}_k$ of R_{l_1} we then have

$$\hat{T}_g \boldsymbol{u}_i = \sum_m D_{mi}(g)\, \boldsymbol{u}_m \tag{4.10}$$

and for the basis vectors $\boldsymbol{v}_k$ of R_{l_2} we have

$$\hat{T}'_g \boldsymbol{v}_k = \sum_n D'_{nk}(g)\, \boldsymbol{v}_n \tag{4.11}$$

Consider a vector $x(x_1, x_2, \ldots, x_{l_1})$, in R_{l_1} and a vector $y(y_1, y_2, \ldots, y_{l_2})$ in R_{l_2}. Let us form $l_1 l_2$ products $x_i y_k$ of the components of x and y, and consider these numbers as the components of a vector in the space $R_{l_1 l_2}$. We shall call this vector the direct product of x and y. The space $R_{l_1 l_2}$ will be called the direct product of R_{l_1} and R_{l_2} and will be denoted by $R_{l_1} \times R_{l_2}$. It is clear that the basis of the space $R_{l_1} \times R_{l_2}$ can be formed from the direct products of the basis vectors u_i and v_k of R_{l_1} and R_{l_2}:

$$w_{ik} = u_i \times v_k \tag{4.12}$$

Let us now define linear operators $\hat{T}''_g$, acting in the space $R_{l_1} \times R_{l_2}$, by the formula

$$\hat{T}''_g w_{ik} = \hat{T}_g u_i \times \hat{T}'_g v_k = \sum_{m,n} D_{mi}(g)\, u_m \times D_{nk}(g)\, v_n$$

$$= \sum_{m,n} D_{mi}(g)\, D'_{nk}(g)\, w_{mn} \tag{4.13}$$

We see that the operators $\hat{T}''_g$ correspond to the direct product of the matrices $D(g)$ and $D'(g)$.

We shall now verify that the matrices $D(g) \times D'(g)$ form a representation of the group G. Assuming that $g_i g_k = g_l$, we have

$$\left.\begin{aligned} D(g_i)\, D(g_k) &= D(g_l) \\ D'(g_i)\, D'(g_k) &= D'(g_l) \end{aligned}\right\} \tag{4.14}$$

and, using property (a) of the direct product of matrices, we obtain

$$(D(g_i) \times D'(g_i))(D(g_k) \times D'(g_k))$$
$$= (D(g_i)\, D(g_k)) \times (D'(g_i)\, D(g_k)) = D(g_l) \times D'(g_l) \tag{4.15}$$

The representation by the matrices $D(g) \times D'(g)$ is called the composition, or the direct product, of the representations $D(g)$ and $D'(g)$. If the representations D and D' are unitary, then by property (b) their direct product is also unitary. If the representations $D^{(i)}$ and $D^{(j)}$ are irreducible, their direct product is, in general, reducible. The decomposition of the direct product of representations into irreducible representations is the Clebsch-Gordan expansion:

$$D^{(i)}(g) \times D^{(j)}(g) = \sum_l^{\oplus} \gamma_{ijl} D^{(l)}(g) \tag{4.16}$$

where $D^{(l)}(g)$ are irreducible representations of the group G. We know that, using the orthogonality relations for the characters of irreducible representations and the known characters of a reducible representation, we can determine how many times it contains each irreducible representation. From (3.88) we have

$$\gamma_{ijs} = \frac{1}{m} \sum_g \bar{\chi}^{(s)}(g) \chi^{(ij)}(g) \tag{4.17}$$

where $\chi^{(ij)}(g)$ denote the characters of the representation $D^{(i)} \times D^{(j)}$. Let us find the expression for the characters $\chi^{(ij)}$. Since the elements of the matrix $D^{(i)} \times D^{(j)}$ are of the form

$$\{D^{(i)} \times D^{(j)}\}_{l\alpha,k\beta} = D^{(i)}_{lk} D^{(j)}_{\alpha\beta} \tag{4.18}$$

it follows that the character $\chi^{(ij)}$ of this representation is given by

$$\chi^{(ij)}(g) = \mathrm{Sp}\left(D^{(i)}(g) \times D^{(j)}(g)\right) = \sum_{l,\alpha} \{D^{(i)}(g) \times D^{(j)}(g)\}_{l\alpha,l\alpha}$$
$$= \sum_{l,\alpha} D^{(i)}_{ll}(g) D^{(j)}_{\alpha\alpha}(g) = \chi^{(i)}(g) \chi^{(j)}(g) \tag{4.19}$$

Thus, the character of the composition of two representations is equal to the product of the characters of the multiplicands. Substituting this result into (4.17), we have

$$\gamma_{ijs} = \frac{1}{m} \sum_g \bar{\chi}^{(s)}(g) \chi^{(i)}(g) \chi^{(j)}(g) \tag{4.20}$$

Suppose that the composition of irreducible representations $D^{(i)}$ and $D^{(j)}$ has been decomposed into two irreducible parts, so that the matrices of this representation have a quasi-diagonal form. We know that this decomposition is obtained as a result of a transformation to a new basis. In our case, this occurs as a result of the transformation from the basis $\boldsymbol{w}_{pk} = \boldsymbol{u}_p^{(i)} \times \boldsymbol{v}_k^{(j)}$ to the basis $\boldsymbol{w}_q^{(s\gamma_s)}$, where s identifies inequivalent irreducible representations and γ_s labels the possibly multiple occurrences of the irreducible representation $D^{(s)}$, the new unit vectors being a linear combination of the old:

$$w_q^{(s\gamma_s)} = \sum_{p,k} (ip, jk \mid s, \gamma_s, q)\, u_p^{(i)} \times v_k^{(j)} \tag{4.21}$$

The coefficients $(ip, jk|s, \gamma_s, q)$ in this formula are Clebsch-Gordan or Wigner coefficients.

If the vectors $w_q^{(s\gamma_s)}$ and $u_p^{(i)} \times v_k^{(j)}$ form an orthonormal basis, the matrix which relates them to the elements $(ip, jk|s, \gamma_s, q)$ should be unitary. When this is so, we have the following orthogonality relations for Clebsch-Gordan coefficients

$$\sum_{p,k} \overline{(ip, jk|s, \gamma_s, q)}(ip, jk|s', \gamma'_s, q') = \delta_{qq'}\delta_{ss'}\delta_{\gamma_s\gamma'_s} \tag{4.22}$$

$$\sum_{s,\gamma_s,q} \overline{(ip, jk|s, \gamma_s, q)}(ip', jk'|s, \gamma_s, q) = \delta_{pp'}\delta_{kk'} \tag{4.23}$$

The composition of three or more representations can be discussed in a similar way.

4.3 Direct product of groups

We shall now introduce the concept of the direct product of groups, and investigate the irreducible representations of the direct product.

Suppose we have two given groups, namely $G^{(1)}$ with elements $g_\alpha^{(1)}$ and $G^{(2)}$ with elements $g_\beta^{(2)}$. We can define a new group $G^{(1)} \times G^{(2)}$, whose elements are the pairs $(g_\alpha^{(1)}, g_\beta^{(2)})$, the order of the elements in the pair being unimportant. This group is the direct product of $G^{(1)}$ and $G^{(2)}$, when the law of 'multiplication' for the group is defined as follows:

$$(g_\alpha^{(1)}, g_\beta^{(2)})(g_{\alpha'}^{(1)}, g_{\beta'}^{(2)}) = (g_{\alpha''}^{(1)}, g_{\beta''}^{(2)}) \tag{4.24}$$

where

$$g_{\alpha''}^{(1)} = g_\alpha^{(1)} g_{\alpha'}^{(1)}, \quad g_{\beta''}^{(2)} = g_\beta^{(2)} g_{\beta'}^{(2)} \tag{4.25}$$

It is readily shown that the unit element of the direct product of two groups is the pair comprising the unit elements of the groups. The element inverse to $(g_\alpha^{(1)}, g_\beta^{(2)})$ is $(g_\alpha^{(1)-1}, g_\beta^{(2)-1})$ An important realization of the direct product of two groups arises when the groups $G^{(1)}$ and $G^{(2)}$ are commuting sub-groups of a given group. In this case, the pair of elements $(g_\alpha^{(1)}, g_\beta^{(2)})$ is interpreted as the result of group multiplication of the elements. We shall show that a direct product cannot be constructed in this way from non-commuting sub-groups. In fact, according to the multiplication rule (4.24), we have

$$(g_\alpha^{(1)}, g_\beta^{(2)})(g_{\alpha'}^{(1)}, g_{\beta'}^{(2)}) = g_\alpha^{(1)} g_{\alpha'}^{(1)} g_\beta^{(2)} g_{\beta'}^{(2)} \tag{4.26}$$

and if we now take the unit elements of the group as the elements $g_\alpha^{(1)}$ and $g_{\beta'}^{(2)}$, we have

$$g_\beta^{(2)} g_{\alpha'}^{(1)} = g_{\alpha'}^{(1)} g_\beta^{(2)} \tag{4.27}$$

which is not satisfied if the sub-groups under consideration do not commute.

We shall show that the number of classes of the group $G^{(1)} \times G^{(2)}$ is equal to the product of the number of classes of the multiplicands $G^{(1)}$ and $G^{(2)}$. To do this, consider a set of those elements of the group $G^{(1)} \times G^{(2)}$, in which the first factor belongs to a given class of $G^{(1)}$, and the second to a given class of $G^{(2)}$ We shall show that these elements of the group $G^{(1)} \times G^{(2)}$ form a class. In point of fact, if the element $(g_\alpha^{(1)}, g_\beta^{(2)})$ belongs to a given set, then for any element $(g_\gamma^{(1)}, g_\delta^{(2)})$ of the group, we have

$$\begin{aligned}&(g_\gamma^{(1)}, g_\delta^{(2)})^{-1}(g_\alpha^{(1)}, g_\beta^{(2)})(g_\gamma^{(1)}, g_\delta^{(2)})\\ &= (g_\gamma^{(1)-1}, g_\delta^{(2)-1})(g_\alpha^{(1)}, g_\beta^{(2)})(g_\gamma^{(1)}, g_\delta^{(2)})\\ &= (g_\gamma^{(1)-1} g_\alpha^{(1)} g_\gamma^{(1)}, g_\delta^{(2)-1} g_\beta^{(2)} g_\delta^{(2)})\end{aligned} \tag{4.28}$$

i.e. we obtain an element belonging to the same set. It is readily seen that all the elements of the set can be obtained as conjugates of any one element. Therefore, the given set of elements does, in fact, form a class of the group $G^{(1)} \times G^{(2)}$, and hence it follows that the number of classes of a direct product is equal to the product of the number of classes of the factors in the direct product.

4.4 Irreducible representations of the direct product of groups

Let us now go on to consider the representations of the direct product of groups. Suppose we are given a representation $D(g_\alpha^{(1)})$ of order l_1 of the group $G^{(1)}$, and a representation $D'(g_\beta^{(2)})$ of order l_2 of the group $G^{(2)}$. We shall show that the direct matrix product $D(g_\alpha^{(1)}) \times D'(g_\beta^{(2)})$ forms representations of order $l_1 l_2$ of the group $G^{(1)} \times G^{(2)}$. In fact, if

$$(g_\alpha^{(1)}, g_\beta^{(2)})(g_{\alpha'}^{(1)}, g_{\beta'}^{(2)}) = (g_{\alpha''}^{(1)}, g_{\beta''}^{(2)}) \tag{4.29}$$

then, according to (4.4), we have for the direct products

$$D(g_\alpha^{(1)}) \times D'(g_\beta^{(2)}) \quad \text{and} \quad D(g_{\alpha'}^{(1)}) \times D'(g_{\beta'}^{(2)})$$

the following equations:

$$\begin{aligned}&(D(g_\alpha^{(1)}) \times D'(g_\beta^{(2)}))(D(g_{\alpha'}^{(1)}) \times D'(g_{\beta'}^{(2)})) \\ &= (D(g_\alpha^{(1)})\,D(g_{\alpha'}^{(1)})) \times (D'(g_\beta^{(2)})\,D'(g_{\beta'}^{(2)})) = D(g_{\alpha''}^{(1)}) \times D'(g_{\beta''}^{(2)})\end{aligned} \tag{4.30}$$

We shall now show that if the representations D and D' are irreducible, the representation $D \times D'$ of the group $G^{(1)} \times G^{(2)}$ is also irreducible. To show this, we shall prove that the only matrix which commutes with all the matrices $D \times D'$ is a multiple of the unit matrix. Let us denote any such matrix by X. We know that the matrix $D \times D'$ can be regarded as a super-matrix whose elements are the matrices $D_{ik}D'$. We shall write down the matrix X in the form of an analogous super-matrix with elements X_{ik} which, in turn, will be matrices of the same order as D'. Super-matrices of the same structure can be multiplied in the same way as ordinary matrices (see Exercise 4.2). We shall thus suppose that the matrix X commutes with all the matrices $D \times D'$. To begin with, let us write down the condition of commutation with all the matrices which correspond to the unit element of the group $G^{(1)}$, taken with any element of $G^{(2)}$, i.e. the matrices $E_{l_1} \times D'(g_\beta^{(2)})$. When written down as super-matrices they have the form of a matrix which is a multiple of the unit matrix and has diagonal elements equal to the matrix D':

$$\begin{pmatrix} X_{11} & X_{12} & \dots & X_{1l_1} \\ \cdot & \cdot & \cdot & \cdot \\ X_{l_11} & X_{l_12} & \dots & X_{l_1l_1} \end{pmatrix} \begin{bmatrix} D' & 0 & 0 & \dots & 0 \\ 0 & D' & 0 & \dots & 0 \\ \cdot & \cdot & \cdot & \cdot & \cdot \\ 0 & 0 & 0 & \dots & D' \end{bmatrix}$$

$$= \begin{bmatrix} D' & 0 & 0 & \dots & 0 \\ 0 & D' & 0 & \dots & 0 \\ \cdot & \cdot & \cdot & \cdot & \cdot \\ 0 & 0 & 0 & \dots & D' \end{bmatrix} \begin{pmatrix} X_{11} & X_{12} & \dots & X_{1l_1} \\ \cdot & \cdot & \cdot & \cdot \\ X_{l_11} & X_{l_12} & \dots & X_{l_1l_1} \end{pmatrix}$$

or

$$\sum_{k=1}^{l_1} X_{ik}\delta_{kj}D' = \sum_{k=1}^{l_1} \delta_{ik}D'X_{kj}$$

and hence

$$X_{ij}D' = D'X_{ij} \tag{4.31}$$

We thus see that all the sub-matrices X_{ij} should commute with all the matrices of the irreducible representation of the group $G^{(2)}$ and, therefore, by Schur's first lemma they are multiples of unit matrices:

$$X_{ik} = x_{ik}E_{l_2} \tag{4.32}$$

where x_{ik} are certain numbers. Let us now write down the condition for X to commute with those matrices of the direct product which correspond to the unit element of the group $G^{(2)}$. These matrices have the form $D \times E_{l_2}$. The commutation condition written in terms of the super-matrices is of the form

$$\sum_k X_{ik} D_{kj} E_{l_2} = \sum_k D_{ik} E_{l_2} X_{kj} \tag{4.33}$$

and hence, using (4.32), we have

$$\sum_k x_{ik} D_{kj} = \sum_k D_{ik} x_{kj} \tag{4.34}$$

We have thus shown that the matrix with elements x_{ik} whose order is l_1 commutes with all the matrices of the irreducible representation $D(g^{(1)})$. Consequently, this matrix is a multiple of the unit matrix and $x_{ik} = \delta_{ik} x$. However, it then follows that the matrix X which is of the form

$$\begin{pmatrix} x_{11}E_{l_2} & x_{12}E_{l_2} & \dots & x_{1l_1}E_{l_2} \\ \cdot & \cdot & \cdot & \cdot \\ x_{l_1 1}E_{l_2} & x_{l_1 2}E_{l_2} & \dots & x_{l_1 l_1}E_{l_2} \end{pmatrix}$$

is also a multiple of the unit matrix. This proves our proposition.

The representations of the group $G^{(1)} \times G^{(2)}$ constructed in this way from the irreducible representations of the groups $G^{(1)}$ and $G^{(2)}$ are not equivalent. This is readily verified using the orthogonality of the characters of these representations which are equal to the products of the characters of the representations D and D'. We shall leave the proof of this to the reader. The number of representations $D \times D'$ is clearly equal to the product of the number of different irreducible representations D and the number of different irreducible representations D', i.e. the product of the number of classes in groups $G^{(1)}$ and $G^{(2)}$. However, since the number of classes of the direct product is equal to the product of the number of classes of the constituent groups, we thus obtain all the irreducible representations of the group $G^{(1)} \times G^{(2)}$.

Exercises

4.1. Prove that the identity representation is contained in the composition of two irreducible representations if, and only if, these representations are complex conjugates.

4.2. Verify that super-matrices with the same division into rows and columns can be multiplied together in accordance with the same rule as for ordinary matrices.

4.3. Show that the irreducible representations of the direct product of two groups constructed in Section 4.3 are not equivalent.

4.4. Show that the sub-groups $\left(g_{\alpha}^{(1)}, E^{(2)}\right)$ and $\left(E^{(1)}, g_{\beta}^{(2)}\right)$ are normal divisors of the group $G^{(1)} \times G^{(2)}$.

4.5. Show that the interchange of the factors in the direct product of matrices is equivalent to a similarity transformation: $A \times B = V(B \times A)V^{-1}$.

Chapter 5

Wigner's Theorem

Having become familiar with some of the basic concepts and theorems of the theory of finite groups, we can go on to consider special sub-groups and applications of the methods of group theory to physical problems. Most of the applications, as we shall see, are based on Wigner's theorem, which will be proved in this chapter.

5.1 The symmetry of a quantum-mechanical system under a group of transformations

We know that the state of a quantum-mechanical system is described by a solution of the Schroedinger equation. Therefore, the symmetry of the system with respect to a given group implies that the corresponding Schroedinger equation is invariant under transformations in this group. We shall confine our attention to the time-independent problem for which the Schroedinger equation is of the form

$$\hat{H}(x)\psi(x) = E\psi(x) \tag{5.1}$$

where x represents the set of variables characterizing the configuration space of the system. We recall (Chapter 3) that the Schroedinger equation is said to be invariant if it preserves its form when the following substitutions are made: $x \to g^{-1}x$, $\psi(x) \to \hat{T}_g\psi(x) = \psi(g^{-1}x)$, where g is a transformation in the symmetry group G of the system. It is clear that the invariance of the Schroedinger equation under the transformation g is a consequence of the invariance of the Hamiltonian of the system:

$$\hat{H}(gx) = H(x) \tag{5.2}$$

We shall show that the invariance of Equation (5.1) under the group G can be written in the form

$$\hat{H}\hat{T}_g = \hat{T}_g\hat{H} \tag{5.2a}$$

which expresses the fact that the operators $\hat{T}_g$ commute with the energy operator $\hat{H}$.

Let $\psi_E(x)$ be an eigenfunction of the operator $\hat{H}$ corresponding to an eigenvalue E. The quantity $\hat{T}_g\psi_E(x)$ is also an eigenfunction of $\hat{H}$ corresponding to the same eigenvalue, i.e.

$$\hat{H}\hat{T}_g\psi_E = E\hat{T}_g\psi_E$$

However,

$$E\hat{T}_g\psi_E = \hat{T}_g\hat{H}\psi_E$$

and, therefore, for any eigenfunction of $\hat{H}$ we have

$$\hat{H}\hat{T}_g\psi_E = \hat{T}_g\hat{H}\psi_E$$

It is clear that this equation is also valid for any function which can be expanded in terms of the eigenfunctions of $\hat{H}$.

The invariance of the Hamiltonian, described by (5.2), can also be written in matrix form. If we use a complete set of orthonormal functions $\psi_i(x)$, we find from (5.2a) that

$$\sum_k H_{ik}D_{kj}(g) = \sum_s D_{is}(g)H_{sj} \tag{5.2b}$$

where $H_{ik} = \int \bar{\psi}_i\hat{H}\psi_k\,dx$, $D_{is} = \int \bar{\psi}_i\hat{T}_g\psi_s\,dx$.

In practice, when we have to solve a quantum-mechanical problem, we frequently have to restrict our attention to an incomplete and non-orthonormal set of functions. We shall show that, in this case, the invariance condition (5.2) retains its form provided a unitary representation of a group G is realized on the chosen set of functions. We shall suppose that for any element g we have

$$\hat{T}_g\psi_i(x)=\psi_i(g^{-1}x)=\sum_k D_{ki}(g)\psi_k(x) \tag{5.3}$$

We shall show that the matrices $D(g)$ form a group representation. Consider two transformations g_1 and g_2, and suppose that $g_1g_2=g_3$. We have, on the one hand,

$$\psi_i\left((g_1g_2)^{-1}x\right)=\psi_i(g_3^{-1}x)=\sum_k D_{ki}(g_3)\psi_k(x) \tag{5.4}$$

and, on the other,

$$\psi_i\left((g_1g_2)^{-1}x\right)=\psi_i(g_2^{-1}g_1^{-1}x)$$

$$=\sum_j D_{ji}(g_2)\psi_j(g_1^{-1}x)=\sum_j D_{ji}(g_2)\sum_k D_{kj}(g_1)\psi_k(x) \tag{5.5}$$

Comparison of (5.4) and (5.5) shows that

$$D_{ki}(g_3)=\sum_j D_{kj}(g_1)D_{ji}(g_2)$$

or

$$D(g_1)D(g_2)=D(g_3)$$

Consider now the matrix

$$H_{ik}=\int\bar{\psi}_i(x)\hat{H}(x)\psi_k(x)\,dx$$

and the analogous matrix for the primed functions $\psi_i'(x)=\psi_i(g^{-1}x)$:

$$H'_{ik}=\int\bar{\psi}_i'(x)\hat{H}(x)\psi_k'(x)\,dx$$

We shall show that since the Hamiltonian is invariant under the transformation g, we must have

$$H'_{ik}=H_{ik} \tag{5.6}$$

In point of fact,

$$H'_{ik}=\int\bar{\psi}_i(g^{-1}x)\hat{H}(x)\psi_k(g^{-1}x)\,dx$$

Substituting $g^{-1}x=x'$ $(dx=dx')$ and using (5.2), we obtain

$$H'_{ik} = \int \bar{\psi}_i(x') H(gx') \psi_k(x') \, dx' = \int \bar{\psi}_i(x) H(x) \psi_k(x) \, dx = H_{ik}$$

On the other hand, according to (5.3),

$$H'_{ik} = \sum_{l,\, j} \int \bar{D}_{li}(g) \bar{\psi}_l(x) H(x) D_{jk}(g) \psi_j(x) \, dx = \sum_{l,\, j} \bar{D}_{li} H_{lj} D_{jk}$$

i.e.

$$H' = D^{+} H D \tag{5.7}$$

If the matrices D are unitary, i.e. $D^{+} = D^{-1}$, we have

$$H' = D^{-1} H D \tag{5.7a}$$

But since $H' = H$, we finally have

$$D(g) H = H D(g) \tag{5.8}$$

which was to be proved.

The symmetry of a quantum-mechanical system under a group of transformations can be expressed as a commutation condition for the Hamiltonian and the matrices of the unitary representation of this group.

5.2 Symmetry of a system of particles executing small oscillations

The energy of a system of N particles executing small oscillations can be written down as the sum of quadratic forms for the kinetic and potential energies:

$$T = \sum_{i=1}^{3N} \frac{m_i \dot{x}_i^2}{2}, \quad V = \frac{1}{2} \sum_{i,\, k=1}^{3N} v_{ik} x_i x_k \tag{5.9}$$

where m_i are the particle masses and x_i are the Cartesian components of their displacements from the equilibrium positions. In classical physics, the quantity $m_i x_i$ is a component of the momentum, whereas in the quantum-mechanical approach it is a component of the momentum operator $i\hbar \frac{\partial}{\partial x_i}$. It is well known that the solution of the problem of small oscillations is considerably simplified by the simultaneous diagonalization of the matrices of these two quadratic forms. The matrix $\|v_{ik}\|$ is a real Hermitian matrix. If we substitute $y_i = \sqrt{m_i}x_i$, we can express the kinetic energy as the sum

$$T = \frac{1}{2}\sum_{i=1}^{3N} \dot{y}_i^2 \tag{5.10}$$

and, consequently, the problem now reduces to the diagonalization of the matrix of the quadratic form corresponding to potential energy, which in terms of the new variables is given by

$$V = \frac{1}{2}\sum_{i,k} \tilde{v}_{ik} y_i y_k, \quad \tilde{v}_{ik} = \frac{v_{ik}}{\sqrt{m_i m_k}} \tag{5.11}$$

The matrix $\|\tilde{v}_{ik}\|$ is again Hermitian. It can therefore be reduced to the diagonal form with the aid of the unitary transformation u:

$$u^{-1}\tilde{v}u = [\lambda_1, \lambda_2, \ldots, \lambda_{3N}]$$

where $\lambda_1, \lambda_2, \ldots, \lambda_{3N}$ are the eigenvalues of the matrix $\tilde{v}$. If we now substitute

$$q_i = \sum_k u_{ik} y_k \tag{5.12}$$

we obtain

$$V = \frac{1}{2}\sum_i \lambda_i q_i^2 \tag{5.13}$$

Because the transformation (5.12) is unitary, the kinetic-energy matrix retains its form:

$$T = \frac{1}{2}\sum_i \dot{q}_i^2 \tag{5.14}$$

If the system is stable, i.e. if the potential energy has a minimum in the equilibrium position, the eigenvalues should be non-negative and can be written in the form

$$\lambda_i = \omega_i^2 \tag{5.15}$$

The Lagrangian, written in terms of the variables q_i,

$$L = T - V = \frac{1}{2}\sum_i (\dot{q}_i^2 - \omega^2 q_i^2) \tag{5.16}$$

leads to the following set of independent equations of motion:

$$\ddot{q}_i + \omega_i^2 q_i = 0 \tag{5.17}$$

Hence it follows that the quantities $q_i(t)$ describe independent oscillators of frequency ω_i. The variables q_i are called the normal coordinates of the system. In the quantum-mechanical description of the system, the Hamiltonian $\hat{H} = T + \hat{V}$, expressed in terms of the normal coordinates, is given by

$$\hat{H} = \frac{1}{2}\sum_{i=1}^{3N}\left(-\hbar^2 \frac{\partial^2}{\partial q_i^2} + \omega^2 q_i^2\right) \tag{5.16a}$$

and the Schroedinger equation reduces to a system of $3N$ independent equations:

$$-\frac{1}{2}\hbar^2 \frac{\partial^2}{\partial q_i^2}\psi(q_i) + \frac{\omega_i^2}{2} q_i^2 \psi(q_i) = E\psi(q_i) \tag{5.17a}$$

Let us now suppose that the equilibrium positions of the oscillating particles form a symmetric configuration: for example, they lie at the corners of a regular hexagon, cube and so on. We must remember that the subscript i identifies not only the various particles, but also the Cartesian components of the displacement of each particle. Consider now a rotation (or rotation plus inversion) which results in the coincidence of the equilibrium position of equivalent particles. The mutual disposition of the particles remains unaltered under this transformation. The set of all such transformations forms the symmetry group G for our system. Let us now fix our attention on a configuration of the particles which is characterized by the displacements $x_1, x_2, \ldots, x_{3N}$. For each particle we shall introduce a Cartesian system with the origin at the equilibrium position. The corresponding axes for all the particles will be assumed to be parallel. If we now consider the set of $3N$ basis vectors $\boldsymbol{e}_i$, we can associate the $3N$-dimensional vector

$$\boldsymbol{x} = x_1\boldsymbol{e}_1 + x_2\boldsymbol{e}_2 + \ldots + x_{3N}\boldsymbol{e}_{3N} \tag{5.18}$$

with the displacements $x_1, x_2, \ldots, x_{3N}$. Let us now introduce a rotation g which results in the coincidence of the equilibrium positions of equivalent particles. This rotation transforms each vector e_i into some other unit vector, or a linear combination of unit vectors. In this way, with each rotation g we can associate a definite operator $\hat{T}_g$, acting in the $3N$-dimensional space:

$$\hat{T}_g e_i = \sum_k D_{ki}(g) e_k \tag{5.19}$$

When $\hat{T}_g$ acts on the vector x, the result is

$$\hat{T}_g x = x' = \sum_i x_i \hat{T}_g e_i = \sum_k x_i D_{ki}(g) e_k$$

or

$$x'_k = \sum_i D_{ki}(g) x_i \tag{5.20}$$

It is clear that, like the matrices $D(g)$, the operators $\hat{T}_g$ form a representation of our group. The configuration $\{x'_1, \ldots, x'_{3N}\}$ is obtained from the configuration $\{x_1, \ldots, x_{3N}\}$ as a result of the rotation g of the entire set of particles following a renumbering corresponding to the inverse rotation g^{-1} (Fig. 1).

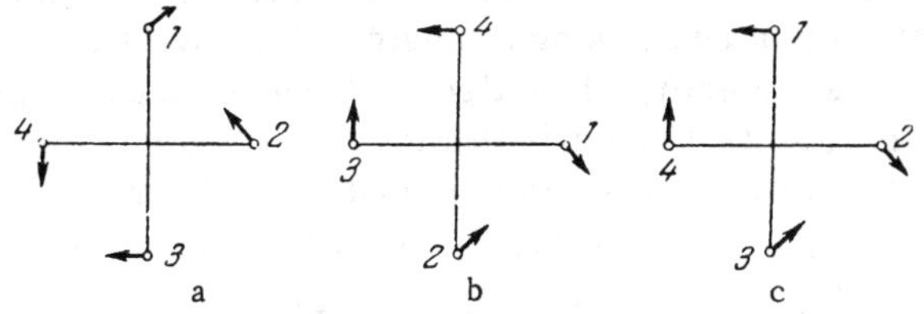

Fig. 1

To establish the explicit form of the matrix $D(g)$, it is convenient to identify the displacements by two indices: $x_{i\alpha}$. The first index denotes the number of the particle and the second the number of the Cartesian component of the displacement. Accordingly, the elements of the matrix D will now be of the form $D_{i\alpha,\, k\beta}(g)$. Let us define the equilibrium position of the i-th particle by the vector $R_i^{(0)}$, and a general position of the same particle by the vector R_i drawn from the centre of mass of the system. The displacement of the

i-th particle can now be represented by

$$\boldsymbol{r}_i = \boldsymbol{R}_i - \boldsymbol{R}_i^{(0)}$$

Let $A(g)$ denote the matrix of the transformation g in three-dimensional space. If the vector $\boldsymbol{R}_k^{(0)}$ is transformed by g into $\boldsymbol{R}_i^{(0)}$, we can write

$$\boldsymbol{r}'_i = \hat{T}_g \boldsymbol{r}_k = A(g)\boldsymbol{R}_k - A(g)\boldsymbol{R}_k^{(0)} = A(g)\boldsymbol{r}_k \tag{5.21}$$

Consequently,

$$D_{i\alpha,\, k\beta}(g) = \delta_{ik} A(g)_{\alpha\beta} \tag{5.22}$$

or

$$D = R(g) \times A(g) \tag{5.22a}$$

where the matrix $R(g)$ consists of zeros and units, and if the equilibrium position of the k-th particle is transformed into the equilibrium position of the i-th particle under the operation g, then

$$R_{ik}(g) = 1, \quad R_{ij}(g) = 0, \quad \text{if} \quad j \neq k$$

Let us now write down the energy for the configuration x'_k of (5.20). To simplify our notation, let us return to our previous numbering, using a single index. We shall suppose that the kinetic energy has already been reduced to the sum of squares, so that the substitution of (5.20), which corresponds to an orthogonal transformation, will preserve its form. The potential energy is then given by

$$V = \frac{1}{2} \sum_{i,j} v_{ij} \sum_k D_{ik} x_k \sum_l D_{jl} x_l$$

$$= \frac{1}{2} \sum_{i,j,k,l} \{D^*\}_{ki} v_{ij} D_{jl} x_k x_l = \frac{1}{2} \sum_{k,l} v'_{kl} x_k x_l$$

where

$$v'_{kl} = \sum_{i,j} \{D^*\}_{ki} v_{ij} D_{jl}$$

The matrix $V' = \| v'_{kl} \|$ can therefore be written in the form

$$V' = D^* V D$$

or, since the matrix D is orthogonal, we have

$$V' = D^{-1}VD \tag{5.23}$$

However, it is clear that the energy corresponding to the configuration $\{x'\}$ should be equal to the energy for the configuration $\{x\}$, since we assume that the energy of the system depends only on the mutual disposition of the particles, which is the same for both configurations. Consequently,

$$\frac{1}{2}\sum_{k,\,l} v'_{kl}x_k x_l = \frac{1}{2}\sum_{k,\,l} v_{kl}x_k x_l$$

and hence $v'_{kl} = v_{kl}$.

The condition that the system should be invariant under the transformation g can, in view of (5.23), be written in the form

$$VD(g) = D(g)V \tag{5.24}$$

i.e. it can be written as a commutation rule for the potential energy matrix and the representation matrices of the symmetry group.

5.3 Wigner's theorem

We shall now prove Wigner's theorem and deduce the consequences of this theorem for the problems considered in Sections 5.1 and 5.2.

Let $D(g)$ be a representation of the group G, which, in general, will be reducible. We shall take the basis elements for this representation so that it splits naturally into irreducible parts. The matrix $D(g)$ will then have the quasi-diagonal form

$$D(g) = \begin{Bmatrix} D^{(1)}(g) & & & \\ & D^{(2)}(g) & & \\ & & D^{(3)}(g) & \\ & & & \ddots \end{Bmatrix}$$

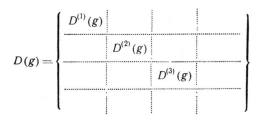

where $D^{(i)}(g)$ is the matrix of the i-th irreducible representation of G. If the representation $D^{(i)}$ appears in the representation D r_i times, we can write (provided the basis vectors of each appearance of $D^{(i)}$ are appropriately chosen)

$$D(g) = \sum_i^{\oplus} E_{r_i} \times D^i(g) \tag{5.25}$$

where E_{r_i} is a unit matrix of order r_i. Next, let us suppose that a matrix H commutes with all the matrices $D(g)$:

$$HD(g) = D(g)H \tag{5.26}$$

We shall show that the matrix H should be of the form

$$H = \sum_i^{\oplus} H^{(i)} \times E_{l_i} \tag{5.27}$$

where $H^{(i)}$ is a matrix of order r_i, E_{l_i} is a unit matrix of order l_i, and l_i is the order of the irreducible representation $D^{(i)}$. The above statement constitutes Wigner's theorem.

Consider a quasi-diagonal matrix D as a diagonal super-matrix with the elements

$$D_{ik} = D^{(i)}\delta_{ik} \tag{5.28}$$

The matrix H will also be represented by a super-matrix with an analogous structure:

$$H = \left\| \begin{array}{ccc|c} H_{11} & H_{12} & H_{13} & \\ \hline H_{21} & H_{22} & H_{23} & \\ \hline H_{31} & H_{32} & H_{33} & \\ \hline & \vdots & \vdots & \cdots \end{array} \right\|$$

Since super-matrices with the same structure can be multiplied together in the same way as ordinary matrices (see

Exercise 4.2), the commutation condition (5.26) can be written in the form

$$\sum_s D_{is}(g) H_{sk} = \sum_s H_{is} D_{sk}(g)$$

or, taking into consideration (5.28),

$$D^{(i)}(g) H_{ik} = H_{ik} D^{(k)}(g) \qquad (5.29)$$

The matrices $D^{(i)}(g)$ and $D^{(k)}(g)$ are the matrices of the irreducible representations of the group G. Therefore, Schur's first and second lemmas lead us to the conclusion that H_{ik} is a zero matrix if $D^{(i)}$ and $D^{(k)}$ are non-equivalent representations, and H_{ik} is a multiple of a unit matrix if the representations $D^{(i)}$ and $D^{(k)}$ coincide. We emphasize that in the last case it is insufficient for $D^{(i)}$ and $D^{(k)}$ to be equivalent; the matrices of these representations should coincide identically (this has already been achieved in (5. 25)).

To determine the explicit form of the matrix H, it will be convenient to introduce the three indices i, ν and α to identify the basis elements for the representation D. The first index i identifies the irreducible representation, the index ν identifies the bases of the equivalent irreducible representations, and the index α identifies the elements of the basis of the irreducible representation. Accordingly, the matrix elements of D and H will carry the six indices: i, ν, α; i', ν' and α'. The elements of H can then be represented by

$$H_{i\nu\alpha;\, i'\nu'\alpha'} = H^{(i)}_{\nu\nu'} \delta_{ii'} \delta_{\alpha\alpha'} \qquad (5.30)$$

from which it follows directly that H can, in fact, be represented by (5.28). For example, if the matrix D is of the form

$$D = \left\{ \begin{array}{c:c:c} D^{(1)} & 0 & 0 \\ \hdashline 0 & D^{(1)} & 0 \\ \hdashline 0 & 0 & D^{(2)} \end{array} \right\}$$

then H will be of the form

$$H = \left\{\begin{array}{cccc:cccc:cccc}
H_{11}^{(1)} & & & & H_{12}^{(1)} & & & & & & & \\
& H_{11}^{(1)} & & & & H_{12}^{(1)} & & & & & & \\
& & \ddots & & & & \ddots & & & 0 & & \\
& & & H_{11}^{(1)} & & & & H_{12}^{(1)} & & & & \\
\hdashline
H_{21}^{(1)} & & & & H_{22}^{(1)} & & & & & & & \\
& H_{21}^{(1)} & & & & H_{22}^{(1)} & & & & & & \\
& & \ddots & & & & \ddots & & & 0 & & \\
& & & H_{21}^{(1)} & & & & H_{22}^{(1)} & & & & \\
\hdashline
& & & & & & & & H_{11}^{(2)} & & & \\
& 0 & & & & 0 & & & & H_{11}^{(2)} & & \\
& & & & & & & & & & \ddots & \\
& & & & & & & & & & & H_{11}^{(2)}
\end{array}\right\}$$

If each irreducible representation of the group G enters the representation D not more than once, then according to (5.30), the matrix H will be diagonal. If, on the other hand, some particular irreducible representation is encountered several times in the expansion of the representation D, then symmetry considerations are insufficient for the complete diagonalization of the matrix H. An additional transformation is necessary in the sub-space of all the basis vectors corresponding to identical irreducible representations. It must be noted, however, that this additional transformation does not change the quasi-diagonal form of the matrix D, and is achieved by diagonalizing the matrices $H^{(i)}$. In point of fact, it is readily verified that a matrix of the form $\sum_i U^{(i)} \times E_{l_i}$, where $U^{(i)}$ is a matrix of order r_i, which diagonalizes the matrix $H^{(i)}$, commutes with the representation matrix:

$$D = \sum_i E_{r_i} \times D^{(i)}$$

If H is completely diagonalized, it can be written in the form

$$H = \sum_i \tilde{H}^{(i)} \times E_{l_i} \tag{5.31}$$

where $\tilde{H}^{(i)}$ are diagonal matrices. If all the eigenvalues of all the matrices $H^{(i)}$ are different, i.e. there is no accidental degeneracy, we may conclude that the eigenvectors of the matrices H belonging to the same eigenvalue are transformed in accordance with one of the irreducible representations of the group G. Consequently, the degree of degeneracy of the eigenvalues of the matrix H should, in this case, be the same as the order of the irreducible representations of the group G.

Consider now some of the consequences which follow from the above theorem.

The above results can be applied directly to the problem of small oscillations. In fact, the symmetry condition for a system executing small oscillations can be expressed as a condition that the potential-energy matrix V commutes with the matrices of the representation $D(g)$ which acts on the displacements. We recall that this form of invariance of the potential energy occurs only because the matrices of the above representation are orthogonal. If we construct symmetrized displacements, i.e. linear combinations of displacements transformed in accordance with the irreducible representations of the group under consideration, the corresponding potential-energy matrix will take the form given by (5.27), i.e.

$$V = \sum_i{}^{\oplus} V^{(i)} \times E_{l_i} \tag{5.32}$$

We see that by constructing symmetrized displacements we simplify the diagonalization of the potential-energy matrix, while in those cases where each irreducible representation is encountered not more than once in the expansion of the representation D this procedure solves the problem completely. If there is no additional degeneracy, then to each eigenfrequency there correspond normal coordinates which transform in accordance with one of the irreducible representations of the group G. The possibility of additional degeneracy depends on accidental coincidence of the eigenvalues of the matrices $V^{(i)}$. By altering the parameters of the problem without affecting its symmetry, we can always ensure that the accidental degeneracy is removed. The fact that the eigenfrequencies of the system correspond to irreducible representations of its symmetry group is usually stated without the qualification that random degeneracy may occur because, in most cases, this degeneracy is absent.

Similar conclusions can be reached for the quantum-mechanical problem which was discussed in Section 5.1. We must, however, emphasize once again that the commutation relation (5.8) is satisfied if, and only if, a unitary representation of the symmetry group under consideration is realized on the chosen set of functions. By taking a complete orthonormal set for our system of functions, we ensure that if there is no accidental degeneracy, then to each eigenvalue of the energy operator there corresponds an irreducible representation by which its eigenfunctions are transformed.

If the basis of the representation is orthonormal and the scalar product is invariant under the group operations, then the representation is unitary. However, the fact that the representation is unitary does not necessarily imply that its basis is orthonormal. Wigner's theorem provides some information about the orthogonality and normalization of the basis elements of a unitary representation D if it is expanded into irreducible parts. Let the basis elements of the reduced representation be denoted by $\varphi_{i\nu\alpha}$, where i, ν and α have the same significance as before, and consider the matrix $\| S_{i\nu\alpha,\, i'\nu'\alpha'} \|$, whose elements are given by

$$S_{i\nu\alpha,\, i'\nu'\alpha'} = (\varphi_{i\nu\alpha},\ \varphi_{i'\nu'\alpha'}) \tag{5.33}$$

where the brackets represent an invariant scalar product. If the representation realized on the elements $\varphi_{i\nu\alpha}$ is unitary, then, as in the case of the Hamiltonian matrix, the invariance condition can be written in the form

$$SD(g) = D(g)S \tag{5.34}$$

and hence, in view of (5.30), we have

$$S_{i\nu\alpha,\, i'\nu'\alpha'} = (\varphi_{i\nu\alpha},\ \varphi_{i'\nu'\alpha'}) = S^{(i)}_{\nu\nu'}\delta_{ii'}\delta_{\alpha\alpha'} \tag{5.35}$$

where $\| S^{(i)}_{\nu\nu'} \|$ is a matrix of order equal to the multiplicity of the i-th irreducible representation in the representation D. The relation given by (5.35) expresses the orthogonality of the basis elements corresponding to non-equivalent representations, and of different elements of the basis of each irreducible representation. We note that the orthogonality of the basis elements of non-equivalent irreducible representations is preserved even for a non-unitary representation.

Chapter 6

Point Groups

Finite sub-groups of the group $\boldsymbol{O}(3)$, i.e. the group of orthogonal transformations in three-dimensional space, are called point groups. In physical applications, point groups are used to describe the symmetry of molecules. Moreover, knowledge of the point groups is necessary for studies in the symmetry properties of crystals. Our immediate ideas about the symmetry of geometrical figures such as prisms, cubes, tetrahedra, etc. depends on the invariance property of these figures under transformations belonging to point groups. In this chapter we shall consider point groups and their irreducible representations. The results will be used to classify electronic and vibrational states of molecules.

6.1 Elements of point groups

The elements of point groups are certain rotations of three-dimensional space, and also rotations accompanied by inversion. We know (see Exercise 1.1) that any element of the rotation group can be represented as a rotation through an angle φ about a certain axis. If the rotation through an angle φ belongs to the group, then rotation through an angle $k\varphi$, where k is an integer, will also belong to the group. Therefore, in a finite group, the angle φ is a rational fraction of 2π. If the smallest angle of rotation about an axis is $\frac{2\pi}{n}$, the axis is called an n-fold axis. The transformation involving rotation through an angle $\frac{2\pi}{n}$ is denoted by C_n or $C_{\mathbf{k}}\left(\frac{2\pi}{n}\right)$, where $\mathbf{k}$ is a unit vector along the axis. It is clear that if the group contains the rotation C_n, it will also contain the rotations C_n^2, C_n^3, ..., C_n^{n-1} through the angles

$$\frac{2\pi}{n}\cdot 2,\ \frac{2\pi}{n}\cdot 3,\ \ldots,\ \frac{2\pi}{n}(n-1) \tag{6.1}$$

These transformations together with the unit element of the group form an Abelian sub-group of the point group.

Let us now consider the elements of point groups containing the inversion i. The inversion transformation converts each vector $\mathbf{r}$ into $-\mathbf{r}$. Since the matrix of this transformation is a multiple of the unit matrix, the transformation will commute with any other orthogonal transformation. Consider the transformation $iC_{\mathbf{k}}(\pi)$. It is clear that the successive application of inversion and rotation through an angle π about a certain axis is equivalent to a mirror reflection in a plane perpendicular to the axis of rotation. Reflection in a plane perpendicular to the vector $\mathbf{k}$ is usually represented by $\sigma_{\mathbf{k}}$. For an arbitrary rotation through an angle φ accompanied by inversion we have

$$iC_{\mathbf{k}}(\varphi) = iC_{\mathbf{k}}(\pi + (\varphi - \pi)) = \sigma_{\mathbf{k}}C_{\mathbf{k}}(\varphi - \pi) \tag{6.2}$$

The transformation $\sigma_{\mathbf{k}}C_{\mathbf{k}}(\varphi)$ is called a mirror rotation and is denoted by $S_{\mathbf{k}}(\varphi)$. The elements of point groups are therefore rotations and mirror rotations.

Let us now consider which elements of a point group can belong to a single class. With this in view, consider the element

$$gC_{\mathbf{k}}(\varphi)g^{-1} \tag{6.3}$$

which is conjugate to $C_{\mathbf{k}}(\varphi)$, where g is an arbitrary element of a point group (i.e. a rotation or mirror rotation). Suppose, to begin with, that g is a rotation and let

$$g\boldsymbol{k}=\boldsymbol{f} \tag{6.4}$$

so that

$$g^{-1}\boldsymbol{f}=\boldsymbol{k} \tag{6.5}$$

The rotation g^{-1} can always be imagined as a rotation $R(\boldsymbol{k}, \boldsymbol{f})$ which maps the vector $\boldsymbol{f}$ onto the vector $\boldsymbol{k}$ (the axis of rotation being perpendicular to the plane of these two vectors), followed by a rotation through an angle α about the vector $\boldsymbol{k}$:

$$g^{-1}=C_{\boldsymbol{k}}(\alpha)R(\boldsymbol{k}, \boldsymbol{f}) \tag{6.6}$$

The element conjugate to $C_{\boldsymbol{k}}(\varphi)$ can therefore be written in the form

$$\begin{aligned} gC_{\boldsymbol{k}}(\varphi)g^{-1} &= R^{-1}(\boldsymbol{k}, \boldsymbol{f})C_{\boldsymbol{k}}^{-1}(\alpha)C_{\boldsymbol{k}}(\varphi)C_{\boldsymbol{k}}(\alpha)R(\boldsymbol{k}, \boldsymbol{f}) \\ &= R^{-1}(\boldsymbol{k}, \boldsymbol{f})C_{\boldsymbol{k}}(\varphi)R(\boldsymbol{k}, \boldsymbol{f}) \end{aligned} \tag{6.7}$$

This transformation can be interpreted as initial mapping of $\boldsymbol{f}$ onto $\boldsymbol{k}$, followed by rotation about $\boldsymbol{k}$ through an angle φ and then by the reverse transition from $\boldsymbol{k}$ to $\boldsymbol{f}$. The final result is therefore rotation through an angle φ about $\boldsymbol{f}$:

$$gC_{\boldsymbol{k}}(\varphi)g^{-1}=C_{\boldsymbol{f}}(\varphi)=C_{g\boldsymbol{k}}(\varphi) \tag{6.8}$$

Similarly, it can be shown that

$$gS_{\boldsymbol{k}}(\varphi)g^{-1}=S_{g\boldsymbol{k}}(\varphi) \tag{6.9}$$

If, on the other hand, g is a mirror rotation, then

$$\left.\begin{aligned} gC_{\boldsymbol{k}}(\varphi)g^{-1} &= C_{-g\boldsymbol{k}}(\varphi) \\ gS_{\boldsymbol{k}}(\varphi)g^{-1} &= S_{-g\boldsymbol{k}}(\varphi) \end{aligned}\right\} \tag{6.10}$$

These relationships lead to the following conclusions.

1. Each class of a point group consists of rotations or mirror rotations through the same angle.

2. Each class contains only those rotations, or mirror rotations through the same angle, for which the axes can be mapped onto each other with the aid of transformations belonging to the group. It must be remembered that, if g is a mirror rotation, the axis $\boldsymbol{k}$ is transformed to $-g\boldsymbol{k}$. It

follows, in particular, that rotations through angles φ and $-\varphi$ about a given axis belong to the same class if the group contains a mirror rotation about this axis, or a reflection plane containing the axis.

6.2 Point groups and their irreducible representations

By establishing the number of classes in a group we simultaneously determine the number of its irreducible representations. Their orders can be found from (3.81). To determine the characters of the irreducible representations we can use the orthogonality relations (3.86) and the equations given by (3.107). The characters of irreducible representations of point groups encountered in applications are listed below. We shall denote the first-order irreducible representations by the letters A and B, the second-order representations by the letter E, and the third-order representations by the letter F. The complex conjugate first-order representations will be combined in pairs, and each pair will also be denoted by the letter E.

I. The group $\boldsymbol{C_n}$

This consists of the rotations C_n^k through angles $\frac{2\pi}{n}k$ about an n-fold axis. The group $\boldsymbol{C_n}$ is one of the realizations of the abstract cyclic group of order n. Since $\boldsymbol{C_n}$ is an Abelian group, all its irreducible representations are of order one. They are determined by the numbers

$$1,\ \varepsilon_l,\ \varepsilon_l^2,\ \ldots,\ \varepsilon_l^{n-1}$$

where $\varepsilon_l = \exp\frac{2\pi i}{n}l$ is the l-th root of the equation $\varepsilon^n = 1$. The characters of irreducible representations of the groups $\boldsymbol{C_2}$, $\boldsymbol{C_3}$ and $\boldsymbol{C_4}$ are as follows:

C_2	E	C_2
A_1	1	1
A_2	1	-1

C_3		E	C_3	C_3^2
A		1	1	1
E		1	ε_1	ε_1^2
		1	ε_1^2	ε_1

C_4		E	C_4	C_4^2	C_4^3
A		1	1	1	1
B		1	-1	1	-1
E		1	i	-1	$-i$
		1	$-i$	1	i

The group $\boldsymbol{C}_2$ is isomorphic to the inversion group $\boldsymbol{i}$. The even (A_1) and odd (A_2) irreducible representations of the latter are denoted by A_g and A_u (where the subscripts g and u are the first letters of the German words gerade and ungerade, i.e. even and odd).

II. The group $\boldsymbol{C}_{nh}$

This group includes the elements of $\boldsymbol{C}_n$ and the reflection σ_h in the plane perpendicular to the axis of rotation. The elements E and σ_h form the sub-group $\boldsymbol{\sigma}_h$ of $\boldsymbol{C}_{nh}$, which is isomorphic to $\boldsymbol{i}$. The Abelian group $\boldsymbol{C}_{nh}$ can be looked upon as the direct product of $\boldsymbol{C}_n$ and $\boldsymbol{\sigma}_h$, i.e. $\boldsymbol{C}_{nh} = \boldsymbol{C}_n \times \boldsymbol{\sigma}_h$. The group $\boldsymbol{C}_{nh}$ has $2n$ first-order irreducible representations. Depending on which of the two irreducible representations of the group σ_h (even or odd) multiplies the representation of the group, the irreducible representations of the group $\boldsymbol{C}_{nh}$ are given the subscripts g and u. For example, for the group $\boldsymbol{C}_{2h}$ we have the following irreducible representations:

C_{2h}	E	C_2	σ_h	$\sigma_h \cdot C_2$
A_g	1	1	1	1
B_g	1	-1	-1	1
A_u	1	1	-1	-1
B_u	1	-1	1	-1

If n is even, then the inversion $i = C(\pi)\sigma_h$ is an element of the group $\boldsymbol{C}_{nh}$ and in this case $\boldsymbol{C}_{nh} = \boldsymbol{C}_n \times \boldsymbol{i}$.

III. The group $\mathbf{C}_{nv}$

In addition to the elements of $\mathbf{C}_n$, this group contains reflections (the operation σ_v) in the n planes passing through the n-fold axis and forming angles $\frac{\pi}{n}$ with one another. This is the symmetry group of a regular n-angle pyramid whose axis is an n-fold axis of symmetry. The group $\mathbf{C}_{nv}$ is not Abelian and it is readily shown that

$$C_n^k \sigma_v = \sigma_v C_n^{-k} \tag{6.11}$$

Let us determine the number of classes of the group $\mathbf{C}_{nv}$. The elements C_n^k and C_n^{-k} belong to a single class and, therefore, if n is even the elements of the sub-group $\mathbf{C}_n$ split into $\frac{n}{2}+1$ classes, or into $\frac{n+1}{2}$ classes if n is odd. All the reflections belong to a single class if n is odd. In this case, all the reflection planes can be obtained from a single plane by rotations through $\frac{2\pi}{n}$. When n is even, rotations through $\frac{2\pi}{n}$ will generate only half of all the reflection planes and, consequently, the reflections split into two classes. Therefore, when n is even, the elements of the group split into $\frac{n}{2}+3$ classes and, when n is odd, into $\frac{n+3}{2}$ classes.

The irreducible representations of the group can be constructed as follows. Consider the basis elements ψ_l ($l = 0, 1, 2, \ldots, n-1$) of the first-order irreducible representations of the group $\mathbf{C}_n$. If we denote the representation operator corresponding to the element C_n^k by $\hat{C}_n^k$, we have

$$\hat{C}_n^k \psi_l = e^{\frac{2\pi k}{n} l} \psi_l \tag{6.12}$$

Moreover, let us define the elements ψ_{-l} by

$$\hat{C}_n^k \psi_{-l} = e^{-\frac{2\pi k i}{n} l} \psi_{-l} \tag{6.13}$$

The elements ψ_l and ψ_{-l} form a two-dimensional invariant sub-space for the group $\mathbf{C}_{nv}$. In fact, using (6.11) it is readily shown that $\hat{\sigma}_v \psi_l = \psi_{-l}$. In view of (6.13) the element ψ_{-l} is linearly independent of ψ_l if $l \neq 0$ and $l \neq \frac{n}{2}$ for even n, and if $l \neq 0$ for odd n. Therefore, in the case of even n we can consider $n-2$ two-dimensional representations of the group C_{nv} which are given by the matrices

$$\left.\begin{aligned} &E \sim \begin{pmatrix} 1 & 0 \\ 0 & 1 \end{pmatrix}, \quad C_n^k \sim \begin{pmatrix} e^{i\frac{2\pi l}{n}k} & 0 \\ 0 & e^{-i\frac{2\pi l}{n}k} \end{pmatrix} \\ &\sigma_v \sim \begin{pmatrix} 0 & 1 \\ 1 & 0 \end{pmatrix}, \quad l = 1, 2, \ldots, \frac{n}{2}-1, \frac{n}{2}+1, \ldots, n-1 \end{aligned}\right\} \qquad (6.14)$$

It is clear that the set of matrices corresponding to the elements C_n^k and C_n^{-k}, which belong to a single class, coincide in the m-th and $(n-m)$-th representations and, consequently, we obtain $\frac{n}{2}-1$ second-order inequivalent representations. The remaining four irreducible representations are of order one.

When n is odd, we obtain $n-1$ two-dimensional irreducible representations from (6.14), of which $\frac{n-1}{2}$ are inequivalent.

Moreover, in this case, we have two first-order irreducible representations. The characters of the irreducible representations of the groups C_{3v} and C_{4v} are given below:

C_{3v}	E	$2C_3$	$3\sigma_v$
A_1	1	1	1
A_2	1	1	−1
E	2	−1	0

C_{4v}	E	C_2	$2C_4$	$2\sigma_v$	$2\sigma_v'$
A_1	1	1	1	1	1
A_2	1	1	1	−1	−1
B_1	1	1	−1	1	−1
B_2	1	1	−1	−1	1
E	2	−2	0	0	0

IV. The group S_{2n}

The elements of this group are the mirror rotations

$S\left(\frac{\pi}{n}\right)=\sigma_n C\left(\frac{\pi}{n}\right)$:

$$E,\ S_{2n},\ S_{2n}^2,\ \ldots,\ S_{2n}^{2n-1} \qquad (6.15)$$

It follows that S_{2n} is a cyclic group of order $2n$. The alternate powers of S_{2n} form a sub-group coinciding with the group C_n.

V. The group D_n

In addition to the elements of the group C_n this group includes rotations through π about n axes perpendicualr to an n-fold axis. This group of rotations maps a regular n-angle prism onto itself. It is readily verified that the group D_n is isomorphic to C_{nv} and, consequently, their irreducible representations coincide.

VI. The group D_{nh}

This is the symmetry group of a regular n-angle prism. Apart from the rotations belonging to D_n it contains the reflection σ_h as well as reflections in the planes passing through its n-fold and two-fold axes. It is a simple matter to write down the characters of this group if we recognize that

$$D_{nh}=D_n\times\sigma_h, \quad \text{if} \quad n=2m+1$$

and

$$D_{nh}=D_n\times i, \quad \text{if} \quad n=2m$$

VII. The group D_{nd}

This is the symmetry group of a body consisting of two regular n-angle prisms placed one on top of the other and rotated through a relative angle of $\frac{\pi}{n}$:

$$D_{2n+1\,d} = D_{2n+1} \times i$$

VIII. The group T *(tetrahedral)*

This group contains all the rotations which map a regular tetrahedron onto itself. Three-fold axes pass through the corners of the tetrahedron (and the center of the opposite faces), while two-fold axes pass through the mid-points of each pair of non-intersecting edges. The twelve elements of the group consist of the unit element, four rotations through $\frac{2\pi}{3}$, four rotations through $\frac{4\pi}{3}$ and three rotations through π, and can be divided into four classes. The group T has four irreducible representations whose orders satisfy (3.81):

$$1^2 + 1^2 + 1^2 + 3^2 = 12$$

T	E	$3C_2$	$4C_3$	$4C_3^2$
A	1	1	1	1
E	1	1	ε	ε^2
	1	1	ε^2	ε
F	3	-1	0	0

where $\varepsilon = e^{\frac{2\pi i}{3}}$

IX. The group $T_h = T \times i$

X. The group T_d

This is the symmetry group of a tetrahedron. It consists of 24 elements divided into five classes. In addition to the rotations belonging to the group T, it contains six reflections in planes passing through two corners and the mid-point of a third side, and two mirror rotations about three two-fold axes. The character of the irreducible representations is as follows:

T_d	E	$8C_3$	$3C_2$	$6\sigma_d$	$6S_4$
A_1	1	1	1	1	1
A_2	1	1	1	−1	−1
E	2	−1	2	0	0
F_1	3	0	−1	1	−1
F_2	3	0	−1	−1	1

*XI. The group **O** (octahedral)*

This group contains all rotations which map a cube onto itself. The 24 elements of this group belong to five classes: the element E, three rotations through π about the axes passing through the mid-points of opposite faces, six rotations through $\pm\frac{\pi}{2}$ about these same axes, six rotations through π about axes passing through the mid-points of opposite edges and eight rotations through $\frac{2\pi}{3}$ about axes passing through opposite corners of the cube. As a result, the group $\boldsymbol{O}$ has five irreducible representations whose orders satisfy the relation

$$1^2+1^2+2^2+3^2+3^2=24$$

The character table for these representations is:

$\boldsymbol{O}$	E	$3C_2$	$6C_4$	$6C_2$	$8C_3$
A_1	1	1	1	1	1
A_2	1	1	−1	−1	1
E	2	2	0	0	−1
F_1	3	−1	1	−1	0
F_2	3	−1	−1	1	0

XII. The group $\boldsymbol{O}_h$

This is the symmetry group of the cube. It is the direct

product of the group $\boldsymbol{O}$ and the inversion group:

$$\boldsymbol{O}_h = \boldsymbol{O} \times \boldsymbol{i} \tag{6.16}$$

The group $\boldsymbol{O}_h$ thus has 48 elements and 10 irreducible representations. Five of these are direct products of the matrices of the irreducible representations of the group $\boldsymbol{O}$ by the matrices of the identity representation of the group $\boldsymbol{i}$. These representations are symmetric under inversion and denoted by $A_1^{(g)}$, $A_2^{(g)}$, $E^{(g)}$, $F_1^{(g)}$, $F_2^{(g)}$ or by Γ_i, $i = 1, 2, \ldots, 5$. The remaining five representations are obtained by multiplying the representations of the group $\boldsymbol{O}$ by the alternating representations of the group $\boldsymbol{i}$. These representations are antisymmetric under inversion and are denoted by $A_1^{(u)}$, $A_2^{(u)}$, $E^{(u)}$, $F_1^{(u)}$, $F_2^{(u)}$ or by Γ_i', $i = 1, 2, \ldots, 5$.

6.3 Classification of normal oscillations and of the electronic states of molecules

We showed in the last chapter that the electronic states of a quantum-mechanical system and the normal coordinates of a system executing small oscillations can be classified in accordance with the irreducible representations of the symmetry group of these systems.

Consider the normal oscillations of a molecule. We shall regard the molecule as a set of particles (nuclei) executing small oscillations about their equilibrium positions which form a symmetric configuration. We note that the normal coordinates of such a system, which correspond to an eigenfrequency, transform in accordance with the irreducible representations of the symmetry group – in our case, the point group of the molecule. The degree of degeneracy of the frequencies is equal to the order of the corresponding irreducible representation. To determine the symmetry properties of the normal coordinates, and the degree of degeneracy of the eigenfrequencies, we must decompose the representation D which governs the transformation of the displacement components x_i into irreducible parts. We know that the number indicating how many times an irreducible representation by the matrices $D^{(i)}$ is contained in a given reducible representation is given by the formula

$$r_i = \frac{1}{m} \sum_g \bar{\chi}(g) \chi^{(i)}(g) \tag{6.17}$$

where $\chi^{(i)}(g)$ is the character of the irreducible representation $D^{(i)}$ of the point group which can be regarded as known, $\chi(g)$ is the character of the reducible representation and m is the order of the group.

The problem is: how can we find the characters of the reducible representation D which is realized on the displacements x_i? It is clear that this can be done by evaluating the traces of the transformation matrices for the displacements $x_1, x_2, \ldots, x_{3N}$ (N is the number of nuclei in the molecule) into the displacements $x'_1, x'_2, \ldots, x'_{3N}$ when the operations g of the symmetry group of the molecule are applied to them. The form of the matrices D was determined in Chapter 5 (see Equation (5.22)). If the position of a given nucleus is transformed by a symmetry transformation into the equilibrium position of another equivalent nucleus, then it is clear that the diagonal elements of the representation matrices corresponding to the displacements of the first nucleus will all be equal to zero. The only non-zero contributions will be due to the nuclei which remain in position under this transformation. At the same time, if the transformation g is a rotation through an angle φ, then the Cartesian components of the displacement of each such nucleus will transform with the aid of the matrix

$$\begin{pmatrix} \cos\varphi & \sin\varphi & 0 \\ -\sin\varphi & \cos\varphi & 0 \\ 0 & 0 & 1 \end{pmatrix} \tag{6.18}$$

whose trace is $1 + 2\cos\varphi$.

For a mirror rotation through an angle φ the transformation matrix is

$$\begin{pmatrix} \cos\varphi & \sin\varphi & 0 \\ -\sin\varphi & \cos\varphi & 0 \\ 0 & 0 & -1 \end{pmatrix} \tag{6.19}$$

and the trace of this matrix is $-1+2\cos\varphi$.

We may thus conclude that the character of the representation corresponding to an element g of a point group is

$$\left.\begin{aligned} \chi(g) &= n_g(1+2\cos\varphi), && g-\text{rotation} \\ \chi(g) &= n_g(-1+2\cos\varphi), && g-\text{mirror rotation} \end{aligned}\right\} \tag{6.20}$$

where n_g is the number of nuclei remaining in position under the transformation g. Substituting these expressions for the characters into Equation (6.17), we can find the reducible representations of the point group transforming the normal coordinates of the molecule. If we know the order of each irreducible representation, we can determine the degree of degeneracy of the eigenfrequencies.

Among the $3N$ degrees of freedom which we have been discussing, three degrees of freedom describe translational motion of the molecule and three the rotation of the molecule as a whole. Since we are considering small displacements of the nuclei, we are concerned with small translational and small rotational motions of the molecule. These degrees of freedom and the corresponding representations are best eliminated from our analysis. The displacements of the nuclei corresponding to these degrees of freedom can be represented by

$$\boldsymbol{r}_i = \boldsymbol{a} \qquad \text{(translation)}$$
$$\boldsymbol{r}_i = [\boldsymbol{\varphi}\boldsymbol{R}_i^{(0)}] \qquad \text{(small rotation)}$$

where $\boldsymbol{a}$ is the translational displacement vector, $\boldsymbol{\varphi}$ is the axial vector representing a small rotation about the axis passing through the origin and $\boldsymbol{R}_i^{(0)}$ defines the position of equilibrium of the i-th nucleus. The components of the vectors $\boldsymbol{a}$ and $\boldsymbol{\varphi}$ can be regarded as the normal coordinates corresponding to translational motion and rotation of the molecule as a whole. Under rotational transformations the two vectors $\boldsymbol{a}$ and φ transform as ordinary three-dimensional vectors. The characters of the corresponding transformation matrices are equal to

$$1 + 2\cos\varphi$$

In the case of mirror rotation, the characters of the transformations of the vectors $\boldsymbol{a}$ and φ are, respectively, equal to $-1 + 2\cos\varphi$ and $1 - 2\cos\varphi$. In view of the foregoing remarks, we have the following expressions for the character of the representation corresponding to the vibrational degree of freedom:

$$\left.\begin{aligned} \chi(g) &= (n_g - 2)(1 + 2\cos\varphi) \quad (g - \text{rotation}) \\ \chi(g) &= n_g(-1 + 2\cos\varphi) \quad (g - \text{mirror rotation}) \end{aligned}\right\} \qquad (6.21)$$

Let us now briefly consider the classification of the electronic states of a molecule. If we write down Schroedinger's equation for the molecule, assuming that the nuclei are fixed at the equilibrium positions, we can show that the eigenfunctions of this equation, i.e. the multi-electron wave functions belonging to a given eigenvalue are transformed by an irreducible representation of the point group of the molecule. The degree of degeneracy of the electronic states should be equal to the order of the irreducible representation. For example, the molecule NH_3 has $\boldsymbol{C}_{3v}$ symmetry and can have only singly or doubly degenerate electronic levels.

We conclude our discussion with this general remark. Any special application of the theory of representations of point groups to the analysis of the electronic states of a molecule requires the introduction of certain approximations. We shall return to this problem in the next chapter.

Exercises

6.1. Determine the types of symmetry of normal coordinates and the degeneracy of the eigenfrequencies for the NH_3 molecule which has $\boldsymbol{C}_{3v}$ symmetry (Fig. 2).

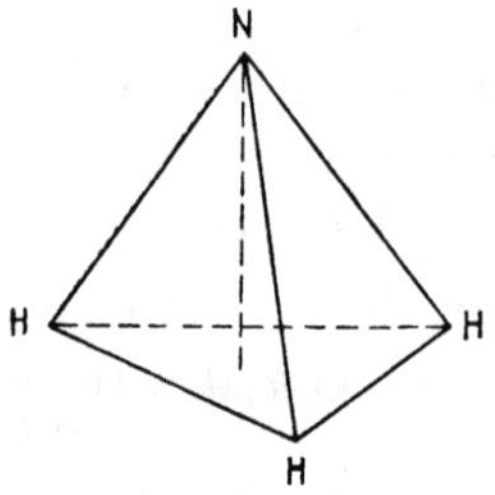

Fig.2

6.2. Write down the character table for the irreducible representations of the group O_h.

Chapter 7

Decomposition of a Reducible Representation into an Irreducible Representation

In the previous chapter we learned how to determine the irreducible representations by which the normal coordinates of a symmetric molecule should transform. However, in many problems it is also necessary to know the explicit expressions for these normal coordinates in terms of the displacements of the nuclei. The determination of the normal coordinates is considerably simplified if we first construct symmetrized displacements from the displacements x_i of the nuclei, i.e. if we find linear combinations of the displacements which transform in accordance with irreducible representations. It is clear that the set of such linear combinations corresponding to one irreducible representation determines an invariant sub-space in the space of the variables x_i. We must therefore decompose the basis space of a certain reducible representation into sub-spaces which are irreducible with respect to the transformations of the symmetry group of the system. An analogous problem arises in quantum mechanics when approximate wave functions of molecules are determined. We shall show how to solve this problem. Our discussion will be general, i.e. it will refer not only to point groups but also to any finite groups.

7.1 Construction of the bases of irreducible representations

Let the set of elements $\{\psi_i\}$ (wave functions or displacements) form the basis of a reducible representation D of a group G. Let us take one element of this basis, say ψ_1, and apply to it all the operations $\hat{T}_g$ corresponding to the transformations g of G. We then obtain a chain of elements transforming one into another. Next, let us take an element out of the set which is not contained in this way, or, more precisely, let us select a basis element which is linearly independent of the elements of the chain. We then construct a second chain from this new element, and so on.

This procedure results in a preliminary decomposition of the representation D. It will not be complete decomposition because the representation realized in the chain will, in general, be reducible. The problem thus reduces to the decomposition of the representations realized in the chains. From the construction of the chains it follows that such representations cannot be broader than the regular representations (Chapter 3). To begin with, let us consider the case where the representation realized in the chain is regular. If we apply all the symmetry operations to the basis element ψ_1 we obtain a chain consisting of m linearly independent elements, where m is the order of G:

$$\hat{T}_g\psi_1 = \psi_g \quad (\psi_1 = \psi_E) \tag{7.1}$$

We know that a regular representation contains all the irreducible representations of G and, moreover, the irreducible representation $D^{(j)}$ of order l_j enters l_j times. Let us find the orthonormal bases which transform in accordance with the representations $D^{(j)}$. To do this, let us construct the operator

$$\hat{P}_{ik}^{(j)} = \sqrt{\frac{l_j}{m}} \sum_g \bar{D}_{ik}^{(j)}(g)\,\hat{T}_g \tag{7.2}$$

where $D_{ik}^{(j)}(g)$ is the matrix element of this representation corresponding to the operation g.

We shall show that for a fixed k the elements

$$\varphi_{ik}^{(j)} = \hat{P}_{ik}^{(j)} \psi_1 = \sqrt{\frac{l_j}{m}} \sum_{g} \bar{D}_{ik}^{(j)}(g) \psi_g \tag{7.3}$$

from a basis of the irreducible representation $D^{(j)}$. In fact, we have

$$\hat{T}_g \varphi_{ik}^{(j)} = \hat{T}_g \sqrt{\frac{l_j}{m}} \sum_{g'} \bar{D}_{ik}^{(j)}(g') \hat{T}_{g'} \psi_1$$

$$= \sqrt{\frac{l_j}{m}} \sum_{g'} \bar{D}_{ik}^{(j)}(g') \hat{T}_g \hat{T}_{g'} \psi_1 = \sqrt{\frac{l_j}{m}} \sum_{g''} \bar{D}_{ik}^{(j)}(g^{-1} g'') \hat{T}_{g''} \psi_1$$

$$= \sqrt{\frac{l_j}{m}} \sum_{g''} \sum_{n} \bar{D}_{in}^{(j)}(g^{-1}) \bar{D}_{nk}^{(j)}(g'') \hat{T}_{g''} \psi_1$$

$$= \sum_{n=1}^{l_j} \bar{D}_{in}^{(j)}(g^{-1}) \sqrt{\frac{l_j}{m}} \sum_{g''} \bar{D}_{nk}^{(j)}(g'') \hat{T}_{g''} \psi_1$$

$$= \sum_{n=1}^{l_j} \bar{D}_{in}^{(j)}(g^{-1}) \varphi_{nk}^{(j)} = \sum_{n=1}^{l_j} D_{ni}^{(j)}(g) \varphi_{nk}^{(j)} \tag{7.4}$$

We see that the operators $\hat{T}_g$ transform the elements $\varphi_{1k}^{(j)}$, $\varphi_{2k}^{(j)}, \ldots, \varphi_{l_j k}^{(j)}$ into one another with matrices $D^{(j)}(g)$. Because of the orthogonality properties of the matrix elements of the irreducible representations, the transformation matrix between ψ and φ in Equation (7.3) is a unitary and, therefore, non-singular matrix, so that the independence of the elements $\varphi_{ik}^{(j)}$ follows from the independence of all the elements of the chain. Moreover, if the elements ψ_g are orthonormal, the elements $\varphi_{ik}^{(j)}$ will also have this property because the transformation is unitary. By assigning the values 1, 2, ..., l_j, to the subscript k, we obtain l_j independent bases which transform in accordance with the representation $D^{(j)}$.

If the number of independent elements ψ_g on the chain is less than m, i.e. if the representation realized on the chain is narrower than the regular representation, then not all the resulting bases of irreducible representations will be independent. The determination of independent bases forms an additional problem. However, the problem is trivial in the case of point groups. The simplification occurs in this case because the orders of the irreducible representations of point groups which are encountered in applications are never greater than 3. If the irreducible representation is one-dimensional it can be encountered not more than once in the representation, and the corresponding basis element is found with the aid of the only possible operator $\hat{P}_{11}^{(j)}$. The two-dimensional irreducible representation can be encountered not more than twice. If it is encountered twice, the discussion given for the regular representation remains valid. If it is encountered once, then the basis of this representation can be obtained by taking any pair of operators $\hat{P}_{11}^{(j)}, \hat{P}_{21}^{(j)}$ or $\hat{P}_{12}^{(j)}, \hat{P}_{22}^{(j)}$ which do not result in zero when applied to the chosen element. The three-dimensional representation can be encountered in D not more than three times. If it is found three times, we must use all three operator triplets

$$\hat{P}_{i1}^{(j)}, \hat{P}_{i2}^{(j)}, \hat{P}_{i3}^{(j)} \qquad (i = 1, 2, 3)$$

If it is found only once, it is sufficient to take the triplet of operators whose application to the element ψ yields a non-zero result. Finally, if the representation is contained twice, then the basis can be obtained by using any two operator triplets yielding linearly independent elements.

We note that in practical applications of this method of decomposition of a reducible representation into an irreducible representation we must know the matrices of the irreducible representations of the corresponding group. (Tables of matrices of irreducible representations of point groups have been given by E. M. Ledovskaya and E. D. Trifonov, Vest. Leningr. Gos. Univ., No 10, p 21 (1962).)

7.2 Determination of the symmetrized displacement of the nuclei of a molecule

To illustrate the above method let us use it to determine the symmetrized displacements of the nuclei of the UF_6 molecule which has the O_h group symmetry (Fig.3). This molecule has 21 degrees of freedom and, therefore, 15 normal coordinates corresponding to vibrations. By using the method given in the previous chapter to determine the characters of a representation D realized on the displacements x_i, y_i, z_i, we obtain the table

E	$3C_2$	$6C_4$	$6C_4^2$	$8C_3$	i	$3C_2i$	$6C_4i$	$6C_4^2i$	$8C_3i$
21	−3	3	−1	0	−3	5	−1	3	0

where x_i, y_i, z_i represent the unit displacements or unit vectors which were denoted by e_i in Chapter 5.

Using (3.59) to determine the structure of the reducible representation and the character table for the irreducible representations of the group O_h (see Section 6.3), we have

$$D = \Gamma_1 + \Gamma_3 + \Gamma_4 + \Gamma_4' + 3\Gamma_5' + \Gamma_5 \qquad (7.5)$$

We note that this decomposition contains representations which transform the coordinates describing the displacements and rotations of the system as a whole.

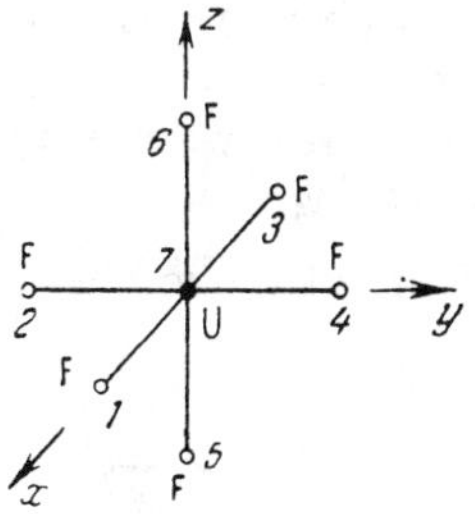

Fig. 3

We also note that the symmetrized displacements for all the irreducible representations other than Γ_5' will, at the same time, be the normal coordinates, since each of the irreducible representations other than Γ_5' enters the decomposition of D not more than once. We are now in a position to find the symmetrized displacements. Instead of considering the entire group O_h we can restrict our attention to the sub-group O if we first replace the displacements x_i, y_i, z_i by their combinations which are symmetric and antisymmetric under inversion:

Symmetric (even) displacements	Antisymmetric (odd) displacements
$x_1 - x_3 \quad x_4 - x_2 \quad x_6 - x_5$	$x_1 + x_3 \quad x_2 + x_4 \quad x_5 + x_6$
$y_1 - y_3 \quad y_4 - y_2 \quad y_5 - y_6$	$y_1 + y_3 \quad y_2 + y_4$
$z_1 - z_3 \quad z_4 - z_2 \quad z_6 - z_5$	$y_5 + y_6 \quad z_1 + z_3$
	$z_2 + z_4 \quad z_5 + z_6$
	$x_7 \quad y_7 \quad z_7$

(7.6)

The basis vectors of the even representations will then be obtained by applying the operators $\hat{P}_{ik}^{(j)}$ to the even displacements, and the basis vectors of the odd representations by applying the operators $\hat{P}_{ik}^{(j)}$ to odd displacements.

A table of the matrices of the irreducible representations of the group O is given in the Appendix at the end of the book.

Let us now apply the operations in the group O to the displacements (7.6). This results in the following five chains:

I. $x_1 - x_3,\ z_6 - z_5,\ y_4 - y_2$
II. $y_1 - y_3,\ z_1 - z_3,\ y_5 - y_6,\ x_4 - x_2,\ z_4 - z_2,\ x_6 - x_5$
III. $x_7,\ y_7,\ z_7$
IV. $y_1 + y_3,\ z_1 + z_3,\ y_5 + y_6,\ x_4 + x_2,\ z_4 + z_2,\ x_5 + x_6$
V. $x_1 + x_3,\ z_6 + z_5,\ y_4 + y_2$

Consider the first of these chains. If we apply the operator

$$\hat{P}_{11}^{(1)} = \frac{1}{24} \sum_g \hat{T}_g \tag{7.7}$$

which corresponds to the identity representation Γ_1 to any of the elements in the first chain, for example. $x_1 - x_3$, we obtain

$$\hat{P}_{11}^{(1)}(x_1 - x_3) = \frac{1}{3}(x_1 - x_3 + z_6 - z_5 + y_4 - y_2) \equiv q_1 \tag{7.8}$$

It was shown in Chapter 5 that the relationship between the displacements $\sqrt{m_i}x_i$ and the coordinates q_k is obtained by a unitary transformation. Therefore, if we have

$$q_i = \sum_k b_{ik} x_k \sqrt{m_k} \tag{7.9}$$

where, for simplicity, the subscript k labels both the atoms and the Cartesian components, then

$$x_k = \frac{1}{\sqrt{m_k}} \sum_i b_{ik} q_i \tag{7.10}$$

This formula gives the displacements of the atoms corresponding to the individual normal coordinates. Assuming that all the coordinates q_i other than q_s are zero, we have

$$\tilde{x}_k = \frac{1}{\sqrt{m_k}} b_{sk} q_s \tag{7.11}$$

Since the atoms corresponding to a chain of displacements are equivalent and, consequently, have equal masses, we can use (7.9) to find the displacements (7.11) (to within the factor $\frac{1}{\sqrt{m}}$). In particular, using (7.8) we find that the normal coordinate q_1 corresponds to the fully symmetric displacement shown in Fig.4.

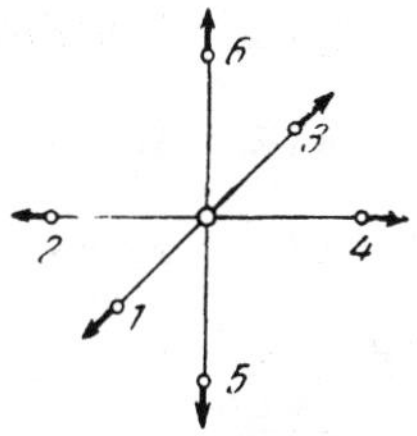

Fig. 4

From the three independent displacements in the first

chain we can construct two further normal coordinates which can only be the basis vectors of the two-dimensional representation Γ_3 since (7.5) contains no further symmetric one-dimensional or two-dimensional irreducible representations. To determine these two normal coordinates let us apply the operators $\hat{P}_{11}^{(3)}$ and $\hat{P}_{21}^{(3)}$ to the displacement $z_6 - z_5$. Using the tables of matrix elements of irreducible representations we have

$$\hat{P}_{11}^{(3)}(z_6 - z_5) = \frac{1}{2}(z_6 - z_5 + y_2 - y_4) = q_2 \tag{7.12}$$

$$P_{21}^{(3)}(z_6 - z_5) = \frac{\sqrt{3}}{6}[2(x_1 - x_3) + z_5 - z_6 + y_2 - y_4] = q_3 \tag{7.13}$$

The displacements corresponding to the normal coordinates q_2 and q_3 are shown in Fig.5.

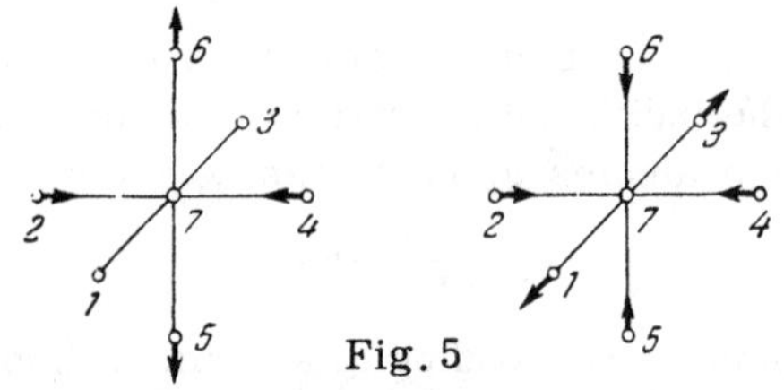

Fig. 5

There remains one further chain (II) on which the symmetric representations Γ_4 and Γ_5 should be realized. To determine the corresponding normal coordinates let us construct the operators $\hat{P}_{i1}^{(4)}$ and $\hat{P}_{i1}^{(5)}$ and apply them to, say, the displacement $y_6 - y_5$:

$$\left.\begin{aligned}\hat{P}_{11}^{(4)}(y_6 - y_5) &= \frac{1}{2}(y_6 - y_5 + z_4 - z_2) = q_4\\ \hat{P}_{21}^{(4)}(y_6 - y_5) &= \frac{1}{2}(x_6 - x_5 + z_1 - z_3) = q_5\\ \hat{P}_{31}^{(4)}(y_6 - y_5) &= \frac{1}{2}(y_1 - y_3 + x_4 - x_2) = q_6\end{aligned}\right\} \tag{7.14}$$

$$\left.\begin{aligned}\hat{P}_{11}^{(5)}(y_6 - y_5) &= \frac{1}{2}(y_6 - y_5 + z_2 - z_4) = q_7\\ \hat{P}_{21}^{(5)}(y_6 - y_5) &= \frac{1}{2}(x_5 - x_6 + z_1 - z_3) = q_8\\ \hat{P}_{31}^{(5)}(y_6 - y_5) &= \frac{1}{2}(y_3 - y_1 + x_4 - x_2) = q_9\end{aligned}\right\} \tag{7.15}$$

The corresponding displacements are shown in Fig.6. The coordinates q_7, q_8, q_9, which transform by the representation Γ_5, describe the rotation of the molecule as a whole.

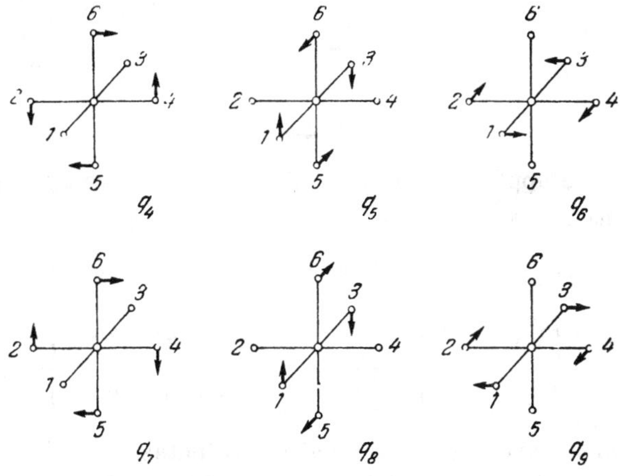

Fig. 6

Let us now consider chains consisting of antisymmetric displacements. We need not apply our method to chain III because any vector transforms in accordance with the irreducible representation Γ_5'. We thus have

$$x_7 = q_{10}, \quad y_7 = q_{11}, \quad z_7 = q_{12} \tag{7.16}$$

The coordinates q_{10}, q_{11}, q_{12} are not normal coordinates because the representation Γ_5' enters the expansion three times. The elements of chain IV transform in accordance with the representations Γ_4' and Γ_5'. If we apply the operators $\hat{P}_{i1}^{(4)}$ to $x_2 + x_4$, we obtain the following normal coordinates:

$$\left.\begin{aligned} \hat{P}_{11}^{(4)}(x_2 + x_4) &= \frac{1}{2}(-x_6 - x_5 + x_2 + x_4) = q_{13} \\ \hat{P}_{21}^{(4)}(x_2 + x_4) &= \frac{1}{2}(y_6 + y_5 - y_1 - y_3) = q_{14} \\ \hat{P}_{31}^{(4)}(x_2 + x_4) &= \frac{1}{2}(z_1 + z_3 - z_4 - z_2) = q_{15} \end{aligned}\right\} \tag{7.17}$$

The corresponding displacements are shown in Fig.7.

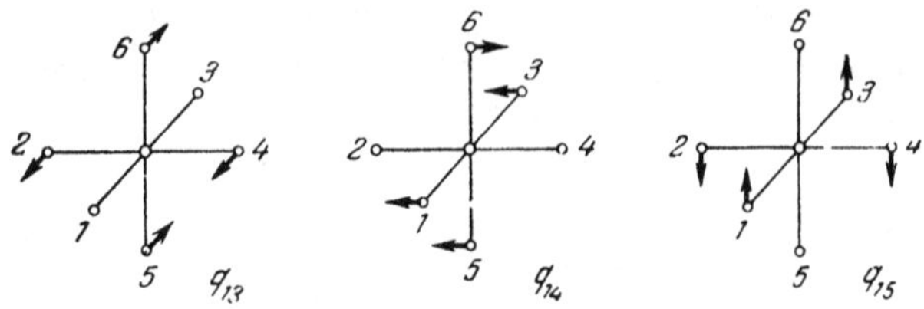

Fig. 7

If we apply the operator $\hat{P}_{11}^{(5)}$ to x_2+x_4, we obtain the following basis elements:

$$\left.\begin{aligned}\hat{P}_{11}^{(5)}(x_2+x_4)&=\frac{1}{2}(x_2+x_4+x_5+x_6)=q_{16}\\ \hat{P}_{21}^{(5)}(x_2+x_4)&=\frac{1}{2}(y_1+y_3+y_5+y_6)=q_{17}\\ \hat{P}_{31}^{(5)}(x_2+x_4)&=\frac{1}{2}(z_4+z_2+z_3+z_1)=q_{18}\end{aligned}\right\} \tag{7.18}$$

which again are not the normal coordinates. Finally, for the last chain (chain V), which consists of three displacements, there remains the irreducible representation Γ_5' and we have the three further coordinates

$$x_1+x_3=q_{19}, \quad z_6+z_5=q_{20}, \quad y_4+y_5=q_{21} \tag{7.19}$$

Out of the nine coordinates formed in accordance with the representation, which is a multiple of the irreducible representation Γ_5', we must separate out the coordinates corresponding to the displacement of the molecule as a whole. This is readily done. It is clear that the three coordinates describing translational motion are

$$\left.\begin{aligned}X&=\frac{\sum m_i x_i}{\sum m_i}\\ Y&=\frac{\sum m_i y_i}{\sum m_i}\\ Z&=\frac{\sum m_i z_i}{\sum m_i}\end{aligned}\right\} \tag{7.20}$$

To construct the normal coordinates out of the remaining six coordinates we must solve the second-order secular equation for the potential-energy matrix (Chapter 5).

7.3 Linear combination of atomic orbitals

As a second example of the application of the method described in Section 7.1 let us consider the electronic states of a molecule.

We shall confine our attention to formulating the problem because its solution is similar to that given in Section 7.2. The adiabatic approximation is often used in analysis of the stationary states of a molecule. In this approximation, the electrons in the molecule are regarded as being located in the electrostatic field of nuclei forming a symmetric configuration. The determination of the state of the electrons in the molecule, i.e. the determination of the eigenfunction of the electronic energy operator for the molecule, can be approximately reduced to a single-electron problem. In this approach each electron is regarded as moving in an effective field due to the remaining electrons and nuclei. It can be assumed approximately that the effective field has the same symmetry as the configuration of the nuclei.

The next step is as follows. If the atoms or ions forming the molecule were at a sufficiently large distance from one another, the single-electron functions of the molecule would be the same as the ionic or atomic single-electron functions. The single-electron levels would then be degenerate, and identical wave functions localized near different but equivalent atoms would correspond to the same value of the energy. In reality, the atoms in the molecule are located at distances at which there is an appreciable interaction between them and, therefore, the single-electron functions for free atoms or ions will not be the solutions of the Schroedinger equation for the single-electron states of the molecule. It can be assumed, however, that the set of such functions forms a system which is sufficiently complete for an approximate solution of our problem. In practice, the number of the functions is finite, and a certain reducible representation D of a point group G is realized on the functions of this set.

It was shown in Chapter 5 that the diagonalization of the Hamiltonian matrix is considerably simplified if we first select our functions to form the bases of irreducible representations. The construction of such bases is analogous to the procedure outlined in Section 7.2. The 'complete' set of functions under consideration is split into chains of functions, and the operators $\hat{P}^{(j)}_{ik}$ are used to construct the bases of irreducible representations in each chain. If a particular irreducible representation is encountered in the expansion only once, the resulting wave functions will be the eigenfunctions of our problem. If the irreducible representation $D^{(j)}$ is encountered r_j times, then after constructing the bases of this irreducible representation to find the eigenfunctions, we must solve the secular equation of order r_j. This method of finding approximate single-electron solutions for a molecular problem is called the LCAO method (linear combination of atomic orbitals).

Exercise

7.1. Construct the wave functions which transform in accordance with the irreducible representations of the group O_h of a cube for a cubic complex such as the F-centre in an alkali halide crystal. It may be assumed that the complete set of functions consists of six s-functions localized around the points *1, 2, ..., 6* (Fig. 8).

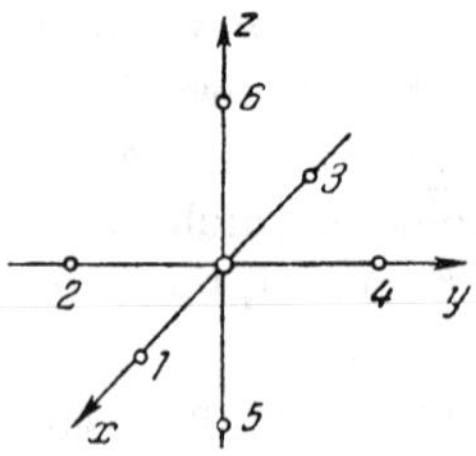

Fig. 8

Chapter 8

Space Groups and Their Irreducible Representations

The symmetry group of an ideal crystal, or the space group, is defined as the set of transformations in three-dimensional space which transform any point of the crystal into an equivalent point. It is important to note that the crystal is either assumed to be infinite, or is replaced by a model in which opposite faces of the specimen are identified. In the latter case the crystal is topologically equivalent to a three-dimensional torus.

8.1 Translation space group

A characteristic feature of the symmetry of crystals which distinguishes them from molecules is the presence of translational symmetry: an ideal crystal is a periodic repetition of a given set of particles. Translational symmetry can be defined with the aid of three non-coplanar vectors $\boldsymbol{a}_1, \boldsymbol{a}_2, \boldsymbol{a}_3$, the basic vectors of the lattice. Translation through the vector

$$\boldsymbol{a} = n_1\boldsymbol{a}_1 + n_2\boldsymbol{a}_2 + n_3\boldsymbol{a}_3 \tag{8.1}$$

where n_1, n_2, n_3 are integers relates the equivalent points $\boldsymbol{r}$

and r' of the crystal:

$$r' = r + a \tag{8.2}$$

The vector a is a lattice vector. If we draw all the vectors a from a given point (the origin), then their end-points will form the Bravais lattice, or 'empty' lattice, corresponding to the given crystal. The end-points of the vectors in this construction are the lattice sites. Three of the basic lattice vectors have the obvious property that the elementary parallelepiped defined by them does not contain a lattice site. We note that the basic vectors cannot be chosen uniquely. However, whatever the choice of these vectors, the volume of the elementary parallelepipeds is always the same. Figure 9 shows how the basic vectors can be chosen for a two-dimensional quadratic lattice. Usually, the basic vectors are chosen to be the shortest of all those possible.

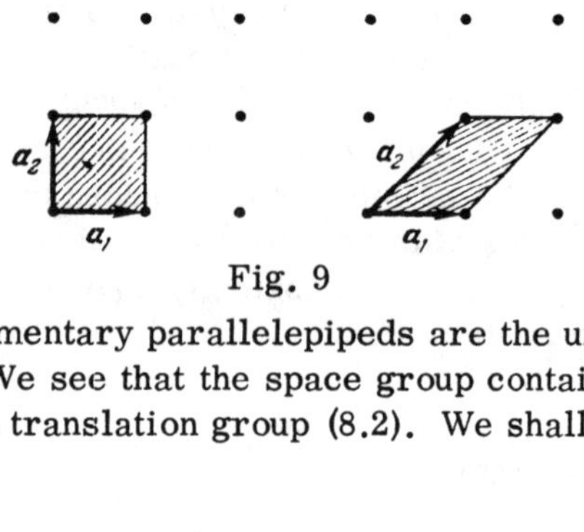

Fig. 9

The elementary parallelepipeds are the unit cells of tne crystal. We see that the space group contains as a sub-group the translation group (8.2). We shall denote it by T_a.

8.2 Syngonies

The Bravais lattices have a definite symmetry with respect to rotations and reflections. For each Bravais lattice there is a point group K whose elements transform each lattice vector into a lattice vector. Orthogonal transformations of three-dimensional space belonging to the group K will be denoted by R. There are seven systems (syngonies) of crystal lattices. It turns out that not all point groups can be lattice symmetry groups. The requirement that both $\boldsymbol{a}$ and $R\boldsymbol{a}$ can simultaneously be lattice vectors restricts the number of possible point groups. Let us now establish these limitations.

We note, to begin with, that the group K should contain inversion: in addition to translation through $\boldsymbol{a}$, the group $T_{\boldsymbol{a}}$ always includes translation through $-\boldsymbol{a}$. To establish the symmetry axes of the group K, let us take the basic lattice vectors $\boldsymbol{a}_1, \boldsymbol{a}_2, \boldsymbol{a}_3$ as the basis unit vectors in the space of the vectors $\boldsymbol{a}$, and write down the transformation R in the new basis in which all the lattice vectors have integral components. If the matrix of the orthogonal transformation R in this basis is denoted by $\tilde{R}$, then

$$\tilde{R} = U^{-1}RU$$

where U is a matrix of the transformation from the initial orthonormal basis to the basis $\boldsymbol{a}_1, \boldsymbol{a}_2, \boldsymbol{a}_3$. If R is a rotation (or mirror rotation) through an angle φ, the traces of the matrices R and $\tilde{R}$ are equal:

$$\operatorname{Sp} \tilde{R} = \operatorname{Sp} R = \pm 1 + 2\cos\varphi \tag{8.3}$$

Since, however, R should transform the lattice vector $\boldsymbol{a}$ into the lattice vector $\boldsymbol{a}' = \tilde{R}\boldsymbol{a}$, it follows that all the elements of $\tilde{R}$, and hence its trace, must be integers. It follows that $\cos\varphi$ can only assume the following values:

$$\cos\varphi = \cos\frac{2\pi n}{m} = \pm 1, \ \pm\frac{1}{2}, \ 0 \tag{8.4}$$

Consequently, the group K can contain only two-, three-, four- and six-fold axes. Finally, it can be shown that if the group K contains the sub-group $\boldsymbol{C}_n$, $n > 2$, it will also contain the sub-group $\boldsymbol{C}_{nv}$. The above three limitations ensure that the point group of a crystal can only be one of the seven point groups

$$\boldsymbol{S}_2, \ \boldsymbol{C}_{2h}, \ \boldsymbol{D}_{2h}, \ \boldsymbol{D}_{3h}, \ \boldsymbol{D}_{4h}, \ \boldsymbol{D}_{6h}, \ \boldsymbol{O}_h$$

This is why there are only seven syngonies: namely, triclinic, monoclinic, rhombic, rhombohedral, tetragonal, hexagonal, and cubic.

The point group of the Bravais lattice imposes definite restrictions on the possible disposition and relative lengths of the basic lattice vectors. We shall not pause to consider this in detail here, but we reproduce below a table giving complete information about the Bravais lattices.

Metric $M_{ik}=(a_i a_k)$	Syngony	Type	Relative length	Relative position
$M_{11} M_{12} M_{13}$ $M_{12} M_{22} M_{23}$ $M_{13} M_{23} M_{33}$	triclinic S_2	simple	any	any
$M_{11} M_{12}$ 0 $M_{12} M_{22}$ 0 0 0 M_{33}	monoclinic C_{2h}	» with centered bases	» »	$a_3 \perp \sigma_{12}$ $c_{32} \perp \sigma_{12}$
M_{11} 0 0 0 M_{22} 0 0 0 M_{33}	rhombic D_{2h}	simple with centered basis body-centered face-centered	» » » »	$a_1 \perp a_2 \perp a_3 \perp a_1$ $c_{32} \perp \sigma_{12} a_1 \perp a_2$ $g_{321} \perp \sigma_{12} a_1 \perp a_2$ $c_{21} \perp a_1 \perp c_{31}$
M 0 0 0 M 0 0 0 M	tetragonal D_{2h}	simple body-centered	$a_1 = a_2$ $a_1 = a_2$	$a_1 \perp a_2 \perp a_3 \perp a_1$ $g_{312} \perp a_2 \perp a_1$
$M \; -\frac{M}{2} \; M_{13}$ $-\frac{M}{2} \; M \; M_{23}$ $M_{13} M_{23} M_{33}$	rhombohedral D_{3d}	simple	$a_1 = a_2$	$g_{312} \perp \sigma_{12}$ $\widehat{a_1 a_2} = \frac{2\pi}{3}$

Metric $M_{ik}=(a_i a_k)$	Syngony	Type	Relative length	Relative position
$\begin{matrix} M & -\frac{M}{2} & 0 \\ -\frac{M}{2} & M & 0 \\ 0 & 0 & M \end{matrix}$	hexa-gonal	simple	$a_1 = a_2$	$a_3 \perp \sigma_{12}$ $\widehat{a_1 a_2} = \frac{2\pi}{3}$
$\begin{matrix} M & 0 & 0 \\ 0 & M & 0 \\ 0 & 0 & M \end{matrix}$	cubic	»	$a_1 = a_2 = a_3$	$a_1 \perp a_2 \perp a_3$
		face-centered	$a_1 = a_2$ $g_{312} = \frac{1}{\sqrt{2}} a_1$	$a_1 \perp a_2 \perp g_{312} \perp a_1$
		body-centered	$a_1 = a_2$ $g_{312} = \frac{1}{2} a_1$	$a_1 \perp a_2 \perp g_{312} \perp a_1$

Notation: $c_{ik} = a_i - \frac{1}{2} a_k$, $g_{ikj} = a_i - \frac{1}{2}(a_k + a_j)$; σ_{ik} is a vector in the plane of the vectors a_i, a_k.

8.3 General element of a space sub-group

The general symmetry element of the empty lattice can be written in the form $t_a R$. The effect of this operation on the three-dimensional vector x is given by the formula

$$t_a R x = R x + a \tag{8.5}$$

It is clear that

$$t_{a_1} R_1 t_{a_2} R_2 x = R_1 R_2 x + a_1 + R_1 a_2 \tag{8.6}$$

$$(t_a R)^{-1} x = R^{-1} x - R^{-1} a \tag{8.7}$$

The operations t_a and R do not commute. We have

$$t_a R = R t_{R^{-1} a} \tag{8.8}$$

We note that the point group K of the lattice is most conveniently defined by considering all operations transforming the vectors $\boldsymbol{a}_1, \boldsymbol{a}_2, \boldsymbol{a}_3, -\boldsymbol{a}_1, -\boldsymbol{a}_2, -\boldsymbol{a}_3$ into one another.

So far, we have been considering the symmetry of empty lattices. Let us return now to the consideration of the symmetry of a crystal.

In addition to the translational sub-group $T_{\boldsymbol{a}}$, the space group contains other transformations whose form depends on the symmetry of the Bravais lattice and the symmetry of the components of the crystal, i.e. on the symmetry of the periodically repeating set of particles forming the crystal. This last fact frequently ensures that not all the transformations in the point group K are included in the symmetry group of the crystal. Not all transformations which map the sites on each other need result in a corresponding mapping of the crystal components. It is therefore possible that the point group of a crystal will only be a sub-group of a point group of an empty lattice.

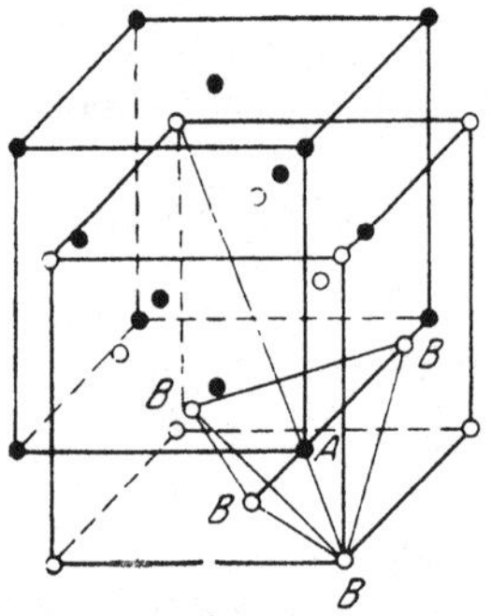

Fig. 10

In general, the symmetry transformation of a crystal which does not contain translations through the lattice vector is of the form $t_{\boldsymbol{\alpha}}R$, where R is a transformation in the point group K and $t_{\boldsymbol{\alpha}}$ is a translation through a vector $\boldsymbol{\alpha}$ which is different from the lattice vector. To explain such non-singular translations, consider the crystal lattice of diamond. The diamond lattice can be constructed from two face-centered cubic lattices displaced relative to each other along the space diagonal of one of the cubes by one-quarter of its length (Fig.10). Let us take the position of any of the nuclei, say the point A, as the center of symmetry. Its nearest neighbors belong to the other sub-lattice. They are denoted by B and form a tetrahedron. It is readily verified that transformations belonging to the group of the tetrahedron will map the diamond lattice onto itself. The inversion transformation i relative to the point A will not be a symmetry element of the lattice. In fact, under this transformation, the nucleus B lying at the corner of the cube is finally found at the mid-point of the space diagonal. However, if the inversion operation is followed by displacement along the diagonal through one-quarter of its length, then the nucleus B will occupy the position of A. It is readily verified that inversion and translation lead to a coincidence of the sub-lattices A and B. This conclusion is also valid for any of the transformations in the set iT_d. It follows that the symmetry elements of the diamond lattice are the transformations of the point group T_d and the transformations of the set iT_d, accompanied by translations t_α, where α is a vector lying along the diagonal of the cube and equal to one-quarter of its length.

We note that the vector α must always be a rational part of the lattice vector $\boldsymbol{a}$. This is so because the product of transformations containing improper translations may result in a simple translation through a lattice vector. The transformations $t_\alpha R$ do not in themselves form a group, since the product of such transformations may contain translations through the lattice vector. If R is a rotation, the group element $t_\alpha R$ is called rotation about the screw axis. If R is a reflection in a plane, the transformation $t_\alpha R$ corresponds to a sliding plane.

The general element of the space group may be written in the form

$$g = t_{\boldsymbol{a}} t_\alpha R = t_{\boldsymbol{a}+\alpha} R \tag{8.9}$$

This is possible in spite of the fact that the translation $t_{\boldsymbol{a}}$ does not commute with $t_\alpha R$. To prove this, we must show that the element $t_a R t_a$ can also be represented by (8.9). In fact,

$$t_a R t_a x = a + R(x + a) = a + Rx + Ra \tag{8.10}$$

Since R belongs to K, it follows that Ra is also a lattice vector and, consequently,

$$t_a R t_a = t_{Ra+a} R \tag{8.11}$$

We have already noted that the transformations $t_a R$ do not form a group. At the same time, it is readily verified that the transformations R do form a group. In fact, we have

$$t_a R t_a \cdot R' = t_a t_{Ra} RR' = t_{a+Ra} RR' \tag{8.12}$$

and hence it follows that in addition to R and R', we always have the transformation RR'. Since $R \in K$, this group should always be a sub-group of K, and we shall denote it by F.

Crystals having the same group F are said to belong to the same class. It is clear that the number of crystal classes is equal to the number of sub-groups of the seven groups $S_2, \ldots, O_h$ defining the syngonies. It is readily shown that these groups contain 32 different sub-groups and, consequently, there are 32 crystal classes. We note that crystals belonging to the same class may belong to different syngonies, whereas crystals belonging to a given class and a given syngony may still differ by the improper translation t_a. It has been found that there are altogether 230 different space groups.

8.4 Irreducible representations of the translation group

To construct the irreducible representations of a space group, let us consider to begin with the irreducible representations of the translation group T_a which is its sub-group.

If translations through the basic lattice vectors are denoted by

$$t_{a_1},\ t_{a_2},\ t_{a_3} \tag{8.13}$$

then an arbitrary element of the group T_a can be represented in the form

$$t_a = t_{a_1}^{n_1} t_{a_2}^{n_2} t_{a_3}^{n_3} \tag{8.14}$$

where

$$a = n_1 a_1 + n_2 a_2 + n_3 a_3 \tag{8.15}$$

All the translations commute with one another and, consequently, the translation group is Abelian. Consider a crystal whose dimensions are determined by the vectors $L_1 a_1$, $L_2 a_2$ and $L_3 a_3$, where L_1, L_2, L_3 are integers. Moreover, let us impose on these translations the so-called cyclic conditions

$$t_{a_i}^{L_i+1} = t_{a_i}, \quad i = 1, 2, 3 \tag{8.16}$$

This is what was meant above by saying that we have identified opposite faces of the specimen.

The translation group T_a can therefore be regarded as the direct product of three cyclic groups T_{a_i} with the elements

$$E, t_{a_i}, t_{a_i}^2, \ldots, t_{a_i}^{L_i-1} \tag{8.17}$$

We know that the irreducible representations of a cyclic group are one-dimensional. For the group T_{a_j} of order L_j they are determined by the numbers

$$e^{2\pi i \frac{m}{L_j} n} \tag{8.18}$$

The index m labels the irreducible representations and can assume the values $0, 1, 2, \ldots, L_j - 1$. The number n is equal to the degree of the corresponding element of the cyclic group (8.17). Therefore, the irreducible representations of the group T_a with the elements (8.14) are the numbers

$$e^{2\pi i \left(\frac{m_1}{L_1} n_1 + \frac{m_2}{L_2} n_2 + \frac{m_3}{L_3} n_3\right)} \tag{8.19}$$

We see that each irreducible representation of the group T_a is determined by the triplet (m_1, m_2, m_3), and the number of different irreducible representations is equal to the product $L_1 L_2 L_3$.

Consider the three vectors b_1, b_2, b_3 defined by

$$(\boldsymbol{a}_l \boldsymbol{b}_k) = 2\pi\delta_{lk}, \qquad l, k = 1, 2, 3 \tag{8.20}$$

It is clear that

$$\boldsymbol{b}_1 = \frac{2\pi\,[\boldsymbol{a}_2\boldsymbol{a}_3]}{(\boldsymbol{a}_1\,[\boldsymbol{a}_2\boldsymbol{a}_3])}, \qquad \boldsymbol{b}_2 = \frac{2\pi\,[\boldsymbol{a}_3\boldsymbol{a}_1]}{(\boldsymbol{a}_2\,[\boldsymbol{a}_3\boldsymbol{a}_1])}, \qquad \boldsymbol{b}_3 = \frac{2\pi\,[\boldsymbol{a}_1\boldsymbol{a}_2]}{(\boldsymbol{a}_3\,[\boldsymbol{a}_1\boldsymbol{a}_2])}$$

Consider the vector

$$\boldsymbol{k} = \frac{m_1}{L_1}\boldsymbol{b}_1 + \frac{m_2}{L_2}\boldsymbol{b}_2 + \frac{m_3}{L_3}\boldsymbol{b}_3 \tag{8.21}$$

The vector $\boldsymbol{k}$ is given in the space defined by the vectors $\boldsymbol{b}_1$, $\boldsymbol{b}_2$, $\boldsymbol{b}_3$ (the reciprocal lattice space with respect to the lattice defined by the vectors $\boldsymbol{a}_1$, $\boldsymbol{a}_2$, $\boldsymbol{a}_3$). It is clear that the numbers (8.19) which give the irreducible representations of the group T_a can be written in the form

$$e^{i(\boldsymbol{k}\boldsymbol{a})} \tag{8.22}$$

Thus, each irreducible representation of the group T_a is characterized by its own vector $\boldsymbol{k}$. We shall denote these representations by $\Gamma_{\boldsymbol{k}}$. If $q_{\boldsymbol{k}}$ is a unit vector of a representation $\Gamma_{\boldsymbol{k}}$, and $\hat{t}_a$ is an operator representing translation through $\boldsymbol{a}$, then

$$\hat{t}_a q_{\boldsymbol{k}} = e^{i(\boldsymbol{k}\boldsymbol{a})} q_{\boldsymbol{k}} \tag{8.23}$$

Two vectors $\boldsymbol{k}$ and $\boldsymbol{k}'$ in the reciprocal lattice space, differing by the vector

$$\boldsymbol{b} = p_1\boldsymbol{b}_1 + p_2\boldsymbol{b}_2 + p_3\boldsymbol{b}_3 \tag{8.24}$$

where p_i are integers, are called equivalent. It is clear that they characterize the same irreducible representation. For the domain of the vector $\boldsymbol{k}$ defining the irreducible representations of the translation group, it is convenient to take a singly connected region in the reciprocal lattice space which contains the origin and satisfies the following two conditions:

a. the region does not contain equivalent vectors;

b. to an arbitrary vector in the reciprocal lattice space one can always find an equivalent vector in this region.

This region is the reduced Brillouin zone.

It may be shown that the groups K_a and K_b of the direct and reciprocal Bravais lattices coincide. In point of fact, let R belong to K_a. In that case, in addition to the equation $(ab) = 2\pi n$, we have $(Ra, b) = 2\pi m$ and hence $(a, R^{-1}b) = 2\pi m$. Consequently, $R^{-1}b$ is a vector of the reciprocal lattice. However, when the transformation R runs over the entire group K_a, the inverse transformation R^{-1} will also run over the entire group. Hence, it may be concluded that $K_a \subset K_b$. If we repeat the discussion for $R \in K_b$, we find that $K_b \subset K_a$. Hence, it follows that the group K_a coincides with K_b.

8.5 Star of the vector *k*

Let us suppose that we know an irreducible representation D of the space group G. In general it will be reducible with respect to the translation sub-group T_a. The representation matrices D corresponding to the translations t_a commute with one another and, therefore, the basis unit vectors of the representation can always be chosen so that these matrices are diagonal. These unit vectors will be the unit vectors of the irreducible representations Γ_k of the translation group, and they can therefore again be denoted by q_k, where k is a vector in the reduced Brillouin zone.

Let us now establish the relationship between the vectors k corresponding to the basis of a given irreducible representation D. Let q_k be one of the unit vectors of this basis, so that from (8.23) we have

$$\hat{t}_a q_k = e^{i(ka)} q_k \tag{8.25}$$

Consider now the operator $\hat{t}_a \hat{t}_\alpha \hat{R}$ corresponding to an arbitrary element of the group G. If we apply this operator to the unit vector q_k, we obtain the unit vector q' given by

$$q' = \hat{t}_a \hat{t}_\alpha \hat{R} q_k \tag{8.26}$$

Let us determine how q' transforms as a result of translations through the lattice vector. We have

$$\hat{t}_{a'} q' = \hat{t}_{a'} \hat{t}_a \hat{t}_\alpha \hat{R} q = \hat{t}_a \hat{t}_\alpha \hat{t}_{a'} \hat{R} q_k \tag{8.27}$$

Using

$$\hat{t}_{a'}\hat{R} = \hat{R}\hat{t}_{R^{-1}a'} \tag{8.28}$$

we have

$$\hat{t}_{a'}q' = \hat{t}_{a'}\hat{t}_{a}\hat{R}\hat{t}_{R^{-1}a'}q_{k} = e^{i(k, R^{-1}a')}\hat{t}_{a'}\hat{t}_{a}\hat{R}q_{k} = e^{i(Rk, a')}q' \tag{8.29}$$

Hence it follows that the vector q' must be given the subscript Rk. Therefore, the general transformation $\hat{t}_a\hat{t}_a\hat{R}$ transforms the unit vector q_k into q_{Rk}. Consequently, if the unit vector q_k is included in the basis of the irreducible representation D of the space group, then it will also contain q_{Rk}, where R is a transformation in the point group F. The set of all inequivalent vectors Rk, where $R \in F$, is the star of the vector k. If all the operations $R \in F$ result in m different vectors

$$k_1 = k, \quad k_2 = R_1 k, \; \ldots, \; k_m = R_{m-1}k$$

then an m-th order star is said to correspond to the vector k.

8.6 The group of the vector k

Let k be a vector in the reduced Brillouin zone. We shall consider all the transformations in G which leave k invariant. They form a sub-group H_k of G. The application of a transformation from G to the vector k must be understood in the sense of (8.29), taking k as the index of the basis vector, i.e. $\hat{t}_a\hat{t}_a\hat{R}k = Rk$. It is clear that all the transformations of the sub-group T_a leave k invariant. If the group G does not contain rotations with improper translations, then apart from the translations t_a, the group H_k will, in addition, include some orthogonal transformations forming a sub-group F_k of the group F, and the products of these transformations with translations. The group H_k will be called the group of the vector k. If the end-point of the vector k lies on the surface of the Brillouin zone, the group H_k will also include elements transforming k to an equivalent vector.

Let us decompose the group G into left cosets with respect to the sub-group H_k:

$$g_1 H_{\boldsymbol{k}},\ g_2 H_{\boldsymbol{k}},\ \ldots,\ g_m H_{\boldsymbol{k}} \tag{8.30}$$

where g_1 is the identity transformation in the group. We shall now establish the properties of the elements g_2, g_3, ..., g_m in this sequence. Since none of the elements in the set $g_2 H_{\boldsymbol{k}}$ should belong to the sub-group $H_{\boldsymbol{k}}$, it follows that the element g_2 cannot leave $\boldsymbol{k}$ invariant. It is readily verified that all the transformations in the coset $g_2 H_{\boldsymbol{k}}$ transform $\boldsymbol{k}$ into the same vector $\boldsymbol{k}_2 = g_2 \boldsymbol{k}$. Similarly, transformations from the other sets transform $\boldsymbol{k}$ into $\boldsymbol{k}_3 = g_3 \boldsymbol{k}$, $\boldsymbol{k}_4 = g_4 \boldsymbol{k}$, ..., $\boldsymbol{k}_m = g_m \boldsymbol{k}$. All these vectors should be different. In fact, if the transformations g_2 and g_3 were to transform $\boldsymbol{k}$ into the same vector $\boldsymbol{k}_2 = \boldsymbol{k}_3$, the set $g_3^{-1} g_2 H_{\boldsymbol{k}} = H_{\boldsymbol{k}}$ should leave $\boldsymbol{k}$ invariant and, consequently, it would be identical with the group $H_{\boldsymbol{k}}$. However, it follows from $g_3^{-1} g_2 H_{\boldsymbol{k}} = H_{\boldsymbol{k}}$ that $g_2 H_{\boldsymbol{k}} = g_3 H_{\boldsymbol{k}}$, which is impossible since cosets do not have common elements. Since the cosets exhaust the entire group, the vectors $\boldsymbol{k}_1$, ..., $\boldsymbol{k}_m$ form the star of $\boldsymbol{k}$. It is readily seen that the group of the vector $\boldsymbol{k}_i$ belonging to this star can be written in the form

$$H_{\boldsymbol{k}_i} = g_i H_{\boldsymbol{k}} g_i^{-1} \tag{8.31}$$

from which it follows that the groups of the vectors of the star are isomorphic to one another.

8.7 Irreducible representations of a space group

Let us suppose that the representation D of a space group has been decomposed into irreducible representations of the translation sub-group, i.e. that the representation matrices corresponding to translations are diagonal. In the space σ of the representation D let us take the basis vectors in which the representation $\Gamma_{\boldsymbol{k}}$ of the group $T_{\boldsymbol{a}}$ is realized. Let us denote by $\sigma_{\boldsymbol{k}}$ the linear sub-space formed by these basis vectors. If we apply a transformation from $H_{\boldsymbol{k}}$ to any vector in the sub-space $\sigma_{\boldsymbol{k}}$, we should again obtain a vector belonging to this sub-space. Hence we may conclude that a representation Γ of the group H_k should be realized in the space σ_k. Similarly, we can isolate sub-spaces σ_{k_i} from the space σ of the representation D in each of which the representations of the group of the corresponding vector are realized. Each of the sub-spaces σ_{k_i} can be obtained from the sub-space σ_k with the aid of the operations g_i. It is also clear that equivalent representations of the isomorphic group H_{k_i}. are realized in each of the sub-spaces σ_{k_i}.

We shall now show that, since the representation D is irreducible, it follows that an irreducible representation of the group of the corresponding vector k_i is realized in each of the sub-spaces σ_{k_i}. Thus, let us suppose that the opposite is the case, i.e. let us suppose that we can isolate a sub-space σ'_k which is invariant with respect to the group H_k in the space σ_k. If we apply the operations $\hat{g}_i\,(i = 1, 2, \ldots, m)$ to σ'_k, we obtain the sub-spaces

$$\sigma'_{k_i} = \hat{g}_i \sigma'_k \tag{8.32}$$

We shall now show that the direct sum of these sub-spaces

$$\sigma' = \sigma'_{k_1} \oplus \sigma'_{k_2} \oplus \ldots \oplus \sigma'_{k_m} \tag{8.33}$$

is then invariant with respect to the entire space group G. According to (8.30), any element of G can be written in the form

$$g = g_i h_k, \quad h_k \in H_k \tag{8.34}$$

in which case we have

$$\hat{g}_j \hat{h}_k \sigma'_{k_i} = \hat{g}_j \hat{h}_k \hat{g}_i \sigma'_k = \hat{g}_l \hat{h}'_k \sigma'_k = \sigma_{k_l} \tag{8.35}$$

We thus see that, under the transformations in G, each of the sub-spaces σ'_k is transformed either into itself or into some other sub-space σ'_k, and the space σ' remains invariant under the transformations in G. On the other hand, σ' is a sub-space of the space σ of the representation D which we have assumed to be irreducible. Hence, it follows that only an irreducible representation of the group H_k can be realized in each sub-space σ_{k_i}, i.e. the representation D_k

is irreducible. We thus conclude that each irreducible representation is determined by the star of the vector $\boldsymbol{k}$ and a certain irreducible representation $\Gamma^{(\alpha)}$ of the group $H_{\boldsymbol{k}}$ of the wave vector. These irreducible representations of a space group will be denoted by $D_{\boldsymbol{k}}^{(\alpha)}$. It is clear that the order $n_{\boldsymbol{k}\alpha}$ of the representation $D_{\boldsymbol{k}}^{(\alpha)}$ is equal to the product of the order n_α of the irreducible representation $\Gamma^{(\alpha)}$ of the group of the wave vector and the number m of vectors in the star $n_{\boldsymbol{k}\alpha} = n_\alpha m$.

8.8 Irreducible representations of the group of the vector $\boldsymbol{k}$

Finally, we must determine which irreducible representations $\Gamma^{(\alpha)}$ of the group $H_{\boldsymbol{k}}$ of the vector $\boldsymbol{k}$ can be realized in the space $\sigma_{\boldsymbol{k}}$. It turns out that certain restrictions must be imposed on the possible irreducible representations of $H_{\boldsymbol{k}}$. In fact, the group of the wave vector includes translations through the lattice vectors. By definition, the entire subspace $\sigma_{\boldsymbol{k}}$ consists of the translation eigenvectors t_a with the eigenvalue $\exp i(\boldsymbol{ka})$. The representation matrix corresponding to translation must therefore be of the form

$$\Gamma^{(\alpha)}(t_a) = e^{i(\boldsymbol{ka})} E_{n_\alpha} \qquad (8.36)$$

where E_{n_α} is the unit matrix. Thus, the only admissible irreducible representations $\Gamma^{(\alpha)}$ of $H_{\boldsymbol{k}}$ will be those in which the matrices (8.36) correspond to translations through the vector $\boldsymbol{a}$. Such irreducible representations of $H_{\boldsymbol{k}}$ will be called normal representations.

Normal irreducible representations can readily be determined when the space group does not contain rotations with improper translations t_α. The group $H_{\boldsymbol{k}}$ then consists of all the possible products of the elements of the group T_a and the point group $F_{\boldsymbol{k}}$, which, in its turn, consists of those elements of the point group F which leave $\boldsymbol{k}$ invariant. Since all the vectors of the space $\sigma_{\boldsymbol{k}}$ are eigenvectors of the translation operations with the same eigenvalue, the irreducibility with respect to the point group F_k follows from the irreducibility of the representation with respect to H_k. The classification of normal irreducible representations of H_k in the case under consideration is thus carried out in accordance with the irreducible representations of the point group F_k.

8.9 Example

Before we go on with the discussion of the space group G containing improper translations, let us illustrate the various concepts introduced above by a simple example of the space group of a monatomic crystal with a square lattice. In this case, the point group F is the group D_4 with eight elements (see Chapter 6). The Brillouin zone will be a square of side $2\frac{\pi}{a}$, where a is the lattice spacing. Since the vector $\boldsymbol{k}$ completely defines the entire star, it is sufficient to consider one-eighth of this square for the classification of the irreducible representations of the space group (Fig. 11).

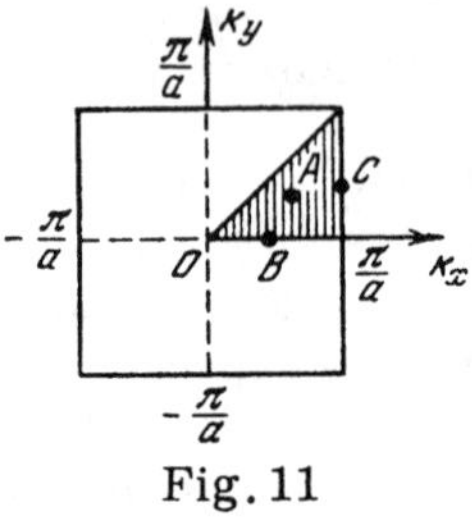

Fig. 11

Consider, to begin with, a vector $\boldsymbol{k}$ which is not connected with the symmetry elements of the Brillouin zone (for example, the vector $\overline{OA}$) in Fig. 11). If we apply to this vector the transformations in D_4, we obtain a star consisting of eight vectors (Fig. 12). It is clear that the group F_k of transformations from D_4 which leave $\boldsymbol{k}$ invariant contains the single element E, and the group H_k coincides in this case with the translation group T_a. The single eighth-order irreducible representation D_{k1} corresponds to the vector $\boldsymbol{k}$. In this representation the matrices corresponding to the translations t_a will be diagonal: their elements will be $\exp i(\boldsymbol{k}_j\boldsymbol{a}),\ j=1,\ 2,\ \ldots,\ 8$ ($\boldsymbol{k}$ are the star vectors). Transformations from the group F permute the basis vectors of the representation and, therefore, each of the matrices corresponding to these transformations has a single non-zero (and non-diagonal) element which is, in fact, equal to unity.

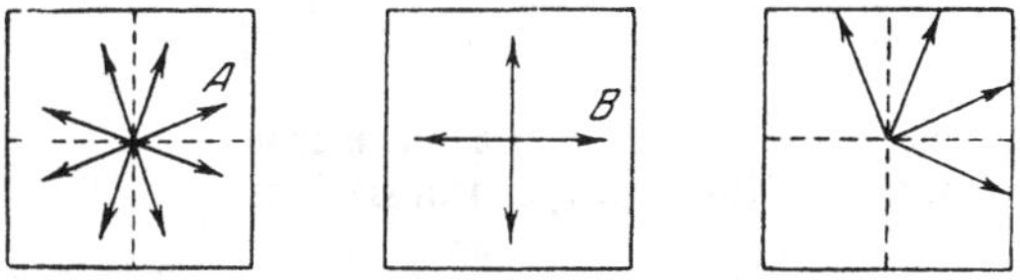

a) b) c)

Fig. 12

Consider now a vector $\boldsymbol{k}$ whose end-point lies on the symmetry axis (for example, $\overline{OB}$ in Fig. 11). In this case, the sub-group F_k contains in addition to the identity element E a further operation σ_y (reflection in the plane XZ) and is isomorphic to the group C_2 having two first-order irreducible representations. The star of the vector $\boldsymbol{k}$ consists of four vectors (Fig. 12b). If we take a vector whose end-point lies on the boundary of the Brillouin zone (for example, the vector $\overline{OC}$ in Fig. 11), we obtain a star consisting of the same four vectors as in Fig. 12b. The remaining four vectors which are obtained by applying transformations from D_4 to $\overline{OC}$ will be equivalent to those shown in the figure. Two irreducible fourth-order representations, $D_k^{(1)}$ and $D_k^{(2)}$, will then correspond to the vector $\boldsymbol{k}$. In these representations the matrices corresponding to translations coincide; the elements of other matrices in these representations can differ in sign only.

Finally, for $\boldsymbol{k}=\boldsymbol{0}$ the group F_k coincides with D_4 and, therefore, the number of different irreducible representations

$D_0^{(\alpha)}$ corresponding to $\boldsymbol{k}=0$ is eoual to their number in the group D_4 (four first-order and one second-order representations). The orders of the representations $D_0^{(\alpha)}$ will be the same as the orders of the representations of the group D_4. The unit matrices correspond to translations $t_{\boldsymbol{a}}$ in the representations $D_0^{(\alpha)}$, and $D_0^{(\alpha)}(R)$ coincide with the corresponding matrices of the irreducible representations of D_4. It is clear that the identity representation of the space group will be one of the irreducible representations corresponding to $\boldsymbol{k}=0$.

8.10 Irreducible representations of a space group containing rotations with improper translations

Consider the case where the group of the vector $\boldsymbol{k}$ contains improper translations. The complication which arises here is connected with the fact that the transformations Rt_α do not form a group: the product of two such elements may contain a translation through the lattice vector. However, we shall see presently that classification of irreducible representations of a group $H_{\boldsymbol{k}}$ for $\boldsymbol{k}$ lying inside the reduced Brillouin zone is also carried out in accordance with irreducible representations of the point group $F_{\boldsymbol{k}}$. A one-to-one correspondence can then be established between the representations of $H_{\boldsymbol{k}}$ and $F_{\boldsymbol{k}}$. Let $\Gamma(h)$ be the representation matrix of the group $H_{\boldsymbol{k}}$, corresponding to the element $h=t_{\boldsymbol{a}}t_\alpha R\in H_{\boldsymbol{k}}$. We shall show that the matrices

$$\tilde{\Gamma}(R)=\Gamma(h)e^{-i[\boldsymbol{k}(\alpha+\boldsymbol{a})]} \tag{8.37}$$

give a representation of the group $F_{\boldsymbol{k}}$. In fact, we have

$$\tilde{\Gamma}(R_1)\tilde{\Gamma}(R_2)=\Gamma(h_1)\Gamma(h_2)e^{-i[\boldsymbol{k}(\boldsymbol{a}_1+\alpha_1)]}e^{-i[\boldsymbol{k}(\boldsymbol{a}_2+\alpha_2)]} \tag{8.38}$$

Moreover,

$$\begin{aligned}\Gamma(h_1h_2)&=\Gamma(t_{\boldsymbol{a}_1}t_{\alpha_1}R_1t_{\boldsymbol{a}_2}t_{\alpha_2}R_2)=\Gamma(t_{\boldsymbol{a}_1}t_{\alpha_1}t_{R\boldsymbol{a}_2}t_{R\alpha_2}R_1R_2)\\&=\tilde{\Gamma}(R_1R_2)e^{i[\boldsymbol{k}(\boldsymbol{a}_1+\alpha_1+R_1\boldsymbol{a}_2+R_1\alpha_2)]}\end{aligned} \tag{8.39}$$

and if we substitute this into (8.38), we have

$$\tilde{\Gamma}(R_1)\tilde{\Gamma}(R_2) = \tilde{\Gamma}(R_1R_2)\,e^{i\,[\boldsymbol{k}\,(R_1\boldsymbol{a}_2 + R_1\alpha_2)] - i\,[\boldsymbol{k}\,(\boldsymbol{a}_2 + \alpha_2)]}$$

$$= \tilde{\Gamma}(R_1R_2)\,e^{i\left(R_1^{-1}\boldsymbol{k} - \boldsymbol{k},\ \boldsymbol{a}_2 + \alpha_2\right)} \tag{8.40}$$

Since by definition the vector $\boldsymbol{k}$ lies inside the reduced Brillouin zone and $R_1 \in F_{\boldsymbol{k}}$, we have

$$R_1^{-1}\boldsymbol{k} = \boldsymbol{k} \tag{8.41}$$

and finally

$$\tilde{\Gamma}(R_1)\tilde{\Gamma}(R_2) = \tilde{\Gamma}(R_1R_2) \tag{8.42}$$

Conversely, it can be shown that we can use (8.37) to construct a representation of the group $H_{\boldsymbol{k}}$ from any representation of $F_{\boldsymbol{k}}$. The existence of this one-to-one correspondence between the representations ensures that if the representation $\tilde{\Gamma}(R)$ is irreducible, then $\Gamma(h)$ is also irreducible. We emphasize once again that the entire foregoing discussion is valid only for a vector $\boldsymbol{k}$ lying inside the Brillouin zone.

Let us now suppose that $\boldsymbol{k}$ lies on the surface of the reduced Brillouin zone. The group $H_{\boldsymbol{k}}$ must then necessarily contain transformations which transform $\boldsymbol{k}$ into a vector which differs from it by the reciprocal lattice vector. The formula given by (8.41) is then no longer valid. If the vector $\boldsymbol{k}$ lies on the surface of the Brillouin zone, it is either equal to a rational fraction of a reciprocal lattice vector, or can be resolved into two vectors, one of which is a rational fraction of a reciprocal lattice vector and the other of which lies inside the Brillouin zone. Let us consider these two cases separately.

Case a. Here $\boldsymbol{k}$ is a rational fraction of a reciprocal lattice vector, and there must be three integers n_1, n_2, n_3 such that

$$\Gamma\left(t_{\boldsymbol{a}_i}^{n_i+1}\right) = E \tag{8.43}$$

(in practice, each of the numbers n_i is not more than 3). This formula is a consequence of the fact that the matrices of the normal representation Γ of $H_{\boldsymbol{k}}$ are of the form

$$\Gamma(t_a) = e^{i(\boldsymbol{ka})}E \tag{8.44}$$

Let us construct the Abelian group $T_{n_1n_2n_3}$, which is homomorphic to the translation group T_a. The elements τ of the group $T_{n_1n_2n_3}$ will be defined by

$$\tau = \tau_1^{s_1}\tau_2^{s_2}\tau_3^{s_3}$$

where s_1, s_2, s_3 assume values between 1 and $n_i + 1$, respectively, where $i = 1, 2, 3$, $\tau_i^{n_i+1} = e$ and e is the unit element of the group. The correspondence between the elements belonging to these groups will be defined as follows:

$$t_{a_i} t_{a_i}^{(n_i+1)l} \to \tau_i \tag{8.45}$$

where l is an arbitrary integer. It is important to note that in the case of (8.43) the normal representations of $H_{\boldsymbol{k}}$ are isomorphic to the representations of $\tilde{H}_{\boldsymbol{k}}$, and consist of all the possible products of $t_u R$ and $\tau_1^{s_1}\tau_2^{s_2}\tau_3^{s_3}$. Therefore, to find the irreducible representations of $H_{\boldsymbol{k}}$ we must find all the irreducible representations of the group $\tilde{H}_{\boldsymbol{k}}$ and select from them those for which the matrices corresponding to the elements τ_1, τ_2, τ_3 are of the form given by (8.44).

Case b. Let us suppose that $\boldsymbol{k} = \boldsymbol{k}_1 + \boldsymbol{k}_2$, where $\boldsymbol{k}_1$ is a rational fraction of a reciprocal lattice vector and $\boldsymbol{k}_2$ lies inside the Brillouin zone. In this case, to find the irreducible representations of $H_{\boldsymbol{k}}$ we must proceed as follows. Let us find the three smallest numbers n_1, n_2, n_3 such that

$$e^{i[\boldsymbol{k}_1 a_j (n_j+1)]} = 1 \tag{8.46}$$

If we now construct a finite group $\tilde{H}_{\boldsymbol{k}_1}$ (homomorphic to $H_{\boldsymbol{k}}$), in which the elements $t_{a_j}(t_{a_j})^{(n_j+1)l}$ correspond to τ_j, we can show that the irreducible representations of $\tilde{H}_{\boldsymbol{k}_1}$ and $H_{\boldsymbol{k}}$ are related by the formula

$$\tilde{\Gamma}(\tau_1^{s_1}\tau_2^{s_2}\tau_3^{s_3}t_u R) = \Gamma(h)\, e^{i[\boldsymbol{k}_2(a+u)]}$$

where the representation $\tilde{\Gamma}$ should have the property that its matrices corresponding to the elements τ_j are of the form given by (8.44).

Exercises

8.1. Show that the translation sub-group is a normal divisor of the space group.

8.2. Show that if the vector $\boldsymbol{k}$ lies inside the reduced Brillouin zone, the vector $R\boldsymbol{k}$, where $R \in K$, also belongs to the reduced zone.

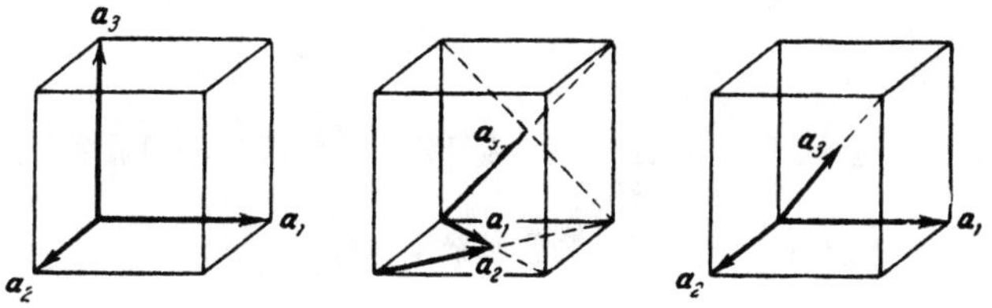

Fig. 13

8.3. Construct the Brillouin zone for simple body- and face-centered cubic lattices. The basic vectors of these lattices are shown in Fig. 13.

Chapter 9

Classification of the Vibrational and Electronic States of a Crystal

We shall now use the irreducible representations of space groups to classify the vibrational and electronic states of a crystal. We shall assume that (1) the normal coordinates of the crystal corresponding to a given frequency transform in accordance with irreducible representations of a space group; (2) the eigenfunctions of the Schroedinger equation for the electron problem, which belong to the same eigenvalue, transform in accordance with the irreducible representations of a space group for fixed equilibrium positions of the nuclei. (In some cases there may be additional degeneracy connected with the invariance under time reversal (Chapter 13). For further details see V. Heine: Theory of Groups in Quantum Mechanics.) For the sake of simplicity we shall confine our attention to space groups which do not contain improper translations.

9.1 Classification of normal vibrations

We shall regard a crystal as a set of particles executing small vibrations about their equilibrium positions. We shall suppose that the equilibrium positions form a configuration having the symmetry of the space group G. We then know (Section 6.3) that the Cartesian components of the particle displacements transform in accordance with an irreducible representation of this group. Let us change from the Cartesian displacements x_i to the normal coordinates q_j (as in Chapter 7, we shall understand x_i to be unit displacements). If x_i is interpreted as a displacement multiplied by the square root of the mass of the corresponding nucleus, then we know that x_i and q_j are related by unitary transformation:

$$x_i = \sum_j c_{ij} q_j \tag{9.1}$$

$$q_j = \sum_i \bar{c}_{ij} x_i \tag{9.2}$$

where $\|c_{ij}\|$ is a unitary matrix. The normal coordinates corresponding to a given frequency must transform in accordance with an irreducible representation of the space group. They can always be chosen so that they will be the eigenvectors of the operator t_a representing translation through a lattice vector. We shall suppose that this has been done, in which case we have from (8.3)

$$\hat{t}_a q_j = e^{i(k_j a)} q_j \tag{9.3}$$

where k_j is a vector defining an irreducible representation of the translation group T_a, by which the normal coordinate q_j is transformed. If, on the other hand, the translation operation is applied to the displacement x_i, we obtain the displacement $x_{i'}$ of an equivalent nucleus in the unit cell, shifted through the vector a relative to the initial cell. By applying the translation operator $\hat{t}_a$ to both sides of (9.1) we therefore have

$$\hat{t}_a x_i = \sum_j c_{ij} \hat{t}_a q_j, \quad x_{i'} = \sum_j c_{ij} e^{i(k_j a)} q_j \tag{9.4}$$

We recall that the quantity q_j is a harmonic function of time

$$q_j = q_j^{(0)} e^{i\omega_j t}$$

We thus see that the displacements $\boldsymbol{x}_i$ form a superposition of harmonic vibrations. The vibrations of equivalent atoms occurring at the same frequency in different cells are periodically shifted in phase in view of (9.4). The wavelength corresponding to each vibration is $\lambda_j = 2\pi|\boldsymbol{k}_j|^{-1}$. In other words, to each normal vibration chosen so that it transforms in accordance with an irreducible representation of the translation group there corresponds a plane wave with a wave vector equal to the vector $\boldsymbol{k}$ of this representation.

The question then is: what is the number of independent waves with the same wave vector $\boldsymbol{k}$? To determine this number we must establish how many times the irreducible representation of a translation group corresponding to this value of $\boldsymbol{k}$ is contained in the representation D which is realized on all the displacements $\boldsymbol{x}_i$. We note that the translation transformation relates the components of the displacements of only the equivalent atoms in different cells of a crystal. If the number of atoms per cell is s, the entire set of displacements $\{\boldsymbol{x}_i\}$ can be split into $3s$ sets, on each of which a regular representation of the translation group is realized. We know that the irreducible representations of the translation group are one-dimensional and, therefore, in the regular representation each of them is encountered only once. It follows that each irreducible representation is encountered precisely $3s$ times in the representation D. Therefore, the answer to the above problem is: the number of different normal vibrations, or the number of different plane waves with the same wave vector, is always equal to $3s$.

Vibrations with $\boldsymbol{k} = \boldsymbol{0}$ are of particular interest among the normal vibrations. Their number is equal to $3s$. It is clear from (9.4) that for these vibrations the motions of the equivalent atoms in different unit cells are in phase. If we regard a crystal as consisting of sub-lattices of equivalent atoms, then $\boldsymbol{k} = \boldsymbol{0}$ corresponds to the vibrations of the sub-lattices relative to each other. These vibrations generally differ both in frequency and in the relative phases of the sub-lattices.

The normal coordinates corresponding to a given value of the wave vector should transform in accordance with an irreducible representation of the group $H_{\boldsymbol{k}}$ of this vector. In the absence of improper translations, the representation Γ is determined by the representation of the point group $F_{\boldsymbol{k}}$ (this is shown in Chapter 8). The normal vibrations which transform in accordance with an irreducible representation of $F_{\boldsymbol{k}}$ will have the same frequency. To find the representation Γ, consider the displacements of atoms belonging to a given fixed cell (the 'zero' cell):

$$\boldsymbol{x}_i^{(0)} = \sum_j c_{ij}\boldsymbol{q}_j, \quad i = 1, 2, \ldots, 3s \tag{9.5}$$

According to (9.4), the displacements of atoms in a unit cell separated from the given cell by the lattice vector $\boldsymbol{a}$ are given by

$$\boldsymbol{x}_i^{(\boldsymbol{a})} = \sum_j c_{ij} e^{i(\boldsymbol{k}_j\boldsymbol{a})}\boldsymbol{q}_j \tag{9.6}$$

Let us consider now certain special displacements $\tilde{\boldsymbol{x}}_i^{(0)}$ and $\tilde{\boldsymbol{x}}_i^{(\boldsymbol{a})}$, which are obtained from (9.5) and (9.6) by setting to zero all the normal coordinates $\boldsymbol{q}_i$ except those for which $\boldsymbol{k}_j = \boldsymbol{k}$. We have

$$\tilde{\boldsymbol{x}}_i^{(0)} = \sum_{\substack{j=1 \\ (\boldsymbol{k}_j = \boldsymbol{k})}}^{3s} c_{ij}\boldsymbol{q}_j \tag{9.5a}$$

$$\tilde{\boldsymbol{x}}_i^{(\boldsymbol{a})} = e^{i(\boldsymbol{k}_j\boldsymbol{a})} \sum_{\substack{j=1 \\ (\boldsymbol{k}_j = \boldsymbol{k})}}^{3s} c_{ij}\boldsymbol{q}_j = e^{i(\boldsymbol{k}_j\boldsymbol{a})}\tilde{\boldsymbol{x}}_i^{(0)} \tag{9.6a}$$

We must now find the transformation rule for the displacements $\tilde{\boldsymbol{x}}_i^{(0)}$. It is clear that the quantities $\boldsymbol{q}_j$ $(\boldsymbol{k}_j = \boldsymbol{k})$ will transform in the same way as $\tilde{\boldsymbol{x}}_i^{(0)}$. In the case of transformations in the point group $F_{\boldsymbol{k}}$, the displacements $\boldsymbol{x}_i^{(0)}$ will transform either into linear combinations of displacements of atoms in the same unit cell, or into linear combinations of displacements of atoms in neighboring cells. However, in view of (9.6a), for the special displacements $\tilde{x}_i^{(0)}$ we can always restrict our attention to linear combinations of displacements of atoms in a given cell. The transformation characters for the quantities $\tilde{x}_i^{(0)}$ can be calculated by generalizing the method introduced in Chapter 6. If as a result of the transformation g in F_k an atom belonging to the zero cell is transformed into an atom of the cell a, the corresponding contribution to the character is

$$\left.\begin{array}{ll}(1+2\cos\varphi)\,e^{i(ka)} & \text{if } g \text{ is a rotation}\\ (-1+2\cos\varphi)\,e^{i(ka)} & \text{if } g \text{ is a mirror rotation}\end{array}\right\} \quad (9.7)$$

The final formulae for the characters of the representation which is realized on the displacements $\tilde{x}_i^{(0)}$ are

$$\chi=(1+2\cos\varphi)\sum_a n_a e^{i(ka)} \quad \text{for rotation} \quad (9.8)$$

$$\chi=(-1+2\cos\varphi)\sum_a n_a e^{i(ka)} \quad \text{for mirror rotation} \quad (9.9)$$

where n_a is the number of atoms in the zero cell which are transformed by the corresponding transformations into the cell a.

It is clear that the representation Γ by which the coordinates q_j with the wave vector $k_j = k$ are transformed has the same characters.

If we now use (3.88), which is based on the orthogonality of the characters of irreducible representations, we obtain the decomposition of the representation Γ into irreducible representations.

We can now summarize our results. Normal oscillations of a crystal are classified with the aid of the wave vector k lying in the Brillouin zone. To each vector k there correspond $3s$ normal coordinates, s being the number of atoms per unit cell. Normal coordinates which transform in accordance with an irreducible representation of the group F_k have the same frequency. The corresponding normal coordinates belonging to other vectors of the star of k have the same frequency.

We have given the above discussion for crystals whose space groups do not contain improper translations. However, all the results can be obtained in a similar way for more complicated groups. The classification of normal coordinates corresponding to a given value of the wave vector is then carried out in accordance with representations of the group H_k and not the group F_k.

We have discussed the normal vibrations of a crystal in terms of symmetry properties only. However, the picture will be incomplete unless we explain some of the additional properties of the frequency spectrum of a crystal. In the harmonic approximation, the displacements of the atoms satisfy the equations

$$\ddot{x}_i^{(a)} = \sum_{j=1}^{3s} \sum_{a} b_{ij}^{(aa')} x_j^{(a')} \tag{9.10}$$

We shall seek the solution corresponding to the contribution of one of the normal vibrations, i.e. we shall set

$$x_j^{(a)} = c_j e^{i(ka) - i\omega t} \tag{9.11}$$

This reduces (9.10) to a set of $3s$ linear homogeneous equations for the coefficients c_i. The condition for a non-trivial solution of the system to exist is that its determinant be zero. We thus obtain an algebraic equation of order $3s$ in ω^2. The solutions of this equation give the eigenfrequencies of the normal vibrations for a given wave vector $\boldsymbol{k}$ whose symmetry we have just determined. The $3s$ roots $\omega_1(\boldsymbol{k})$, $\omega_2(\boldsymbol{k}), \ldots, \omega_{3s}(\boldsymbol{k})$ are the branches of the elastic spectrum. The values of these functions for $\boldsymbol{k}=0$ are the limiting frequencies. We have already shown that, when $\boldsymbol{k}=0$, there are three normal coordinates which describe the displacement of the crystal as a whole (translational degrees of freedom). It is clear that these coordinates correspond to zero frequencies, and we can therefore conclude that three of the spectral branches should begin with $\omega=0$. These are the acoustic branches; the remaining $3s-3$ branches are the optical branches.

We have shown above that the normal coordinates corresponding to a given wave vector and transforming in accordance with an irreducible representation of the group of $\boldsymbol{k}$ correspond to the same frequency. Therefore, it follows that if the order of an irreducible representation of the group of $\boldsymbol{k}$ is greater than unity, there is frequency degeneracy at the given point $\boldsymbol{k}$ in the Brillouin zone, and the branches are said to coincide. This can occur both at individual symmetric points and along the symmetry axes of the Brillouin zone.

9.2 Classification of the electronic states of a crystal

Consider now the electronic states of a crystal, assuming that the nuclei are fixed at the lattice sites. The Hamiltonian for the multi-electron system in the field of these nuclei is invariant under transformations corresponding to a space group. The eigenfunctions of this operator corresponding to the same eigenvalue will therefore transform in accordance with an irreducible representation $D_{\boldsymbol{k}}^{(\alpha)}$ of the symmetry group of the crystal, and can be chosen so that they will be the eigenfunctions of the operator resulting in translations through the lattice vector. We shall denote such functions by $\Psi_{\boldsymbol{k}\alpha j}(r_1, \ldots, r_N)$ and the energy eigenvalues by $E_{\boldsymbol{k}\alpha}$. The subscript $\boldsymbol{k}$ in these quantities represents one of the vectors in the star of an irreducible representation, the symbol α labels the irreducible representations of the wave-vector group, and the symbol j labels the basis vectors of the α representation. We have

$$\hat{t}_a \Psi_{\boldsymbol{k}\alpha j}(r_1, \ldots, r_N) = \Psi_{\boldsymbol{k}\alpha j}(\boldsymbol{r}_1 + \boldsymbol{a}, \ldots, \boldsymbol{r}_N + \boldsymbol{a})$$
$$= e^{i(\boldsymbol{k}\boldsymbol{a})}\Psi_{\boldsymbol{k}\alpha j}(\boldsymbol{r}_1, \ldots, \boldsymbol{r}_N) \tag{9.12}$$

from which it follows that the function

$$U_{\boldsymbol{k}\alpha j}(\boldsymbol{r}_1, \ldots, \boldsymbol{r}_N) = e^{i(\boldsymbol{k}\boldsymbol{R})}\Psi_{\boldsymbol{k}\alpha j}(\boldsymbol{r}_1, \ldots, \boldsymbol{r}_N) \tag{9.13}$$

where $R = \frac{1}{N}\sum_{i=1}^{N} r_i$ is invariant under translations. In fact,

$$\begin{aligned}\hat{t}_a U_k(r_1, \ldots, r_N) &= U_k(r_1 + a, \ldots, r_N + a)\\ &= e^{-i(k, R+a)}\Psi_k(r_1 + a, \ldots, r_N + a)\\ &= e^{-i(k, R+a)}e^{ika}\Psi_k(r_1, \ldots, r_N)\\ &= e^{-i(kR)}\Psi_k(r_1, \ldots, r_N) = U_k(r_1, \ldots, r_N)\end{aligned}$$

and hence it follows that the eigenfunctions of our problem can always be written in the form

$$\Psi_{k\alpha j}(r_1, \ldots, r_N) = e^{i(kR)}U_{k\alpha j}(r_1, \ldots, r_N) \qquad (9.14)$$

where $U_{k\alpha j}(r_1, \ldots, r_N)$ is a periodic function. The functions $U_{k\alpha j}$ corresponding to the same energy eigenvalue transform in accordance with an irreducible representation of the group of the wave vector k. If now the function $U_{k\alpha j}$ is represented by a series expansion in terms of a complete set of periodic functions, and if we substitute $\Psi_{k\alpha j}(r_1, \ldots, r_N)$ defined by (9.14) into the Schroedinger equation, we obtain a set of linear equations for the expansion coefficients. The roots of the corresponding secular equation give the energy eigenvalues as functions of the wave vector:

$$E_1(k),\ E_2(k),\ \ldots$$

Each of these functions is said to define an energy band in the crystal. The values E_l of these functions may be equal at symmetric points in the Brillouin zone (on symmetry axes or planes). The energy bands are then said to coincide, and this is due to the energy degeneracy for the given value of k which we have already mentioned.

9.3 The one-electron approximation

In practice, group methods are applied to simplified models of physical systems. The success of these approximations depends largely on the extent to which the symmetry properties of the exact solution have been taken into account.

The self-consistent field method, in which the problem of interacting electrons is reduced to a one-electron problem, is the most important approximate method. The interaction between electrons is replaced approximately by an effective field having the symmetry of the crystal. The eigenfunction of the one-electron energy operator which has the symmetry group of the crystal can then be written in the form

$$\Psi_{kaj}(\boldsymbol{r}) = e^{i(\boldsymbol{kr})} U_{kaj}(\boldsymbol{r}) \tag{9.15}$$

where

$$U_{kaj}(\boldsymbol{r} + \boldsymbol{a}) = U_{kaj}(\boldsymbol{r}) \tag{9.16}$$

The wave functions (9.15), which are the basis functions of certain irreducible representations of a space group, are called Bloch functions. They can be regarded as generalized plane waves with variable periodic amplitude $U_{\boldsymbol{k}}(\boldsymbol{r})$. The vector $\boldsymbol{k}$ is the crystal momentum of the electron. The energy eigenvalue $E = E(\boldsymbol{k})$ determines the one-electron energy band.

One of the possible methods of solving the one-electron problem is to expand the function Ψ_{kaj} in terms of the one-electron wave functions of atoms or ions forming the crystal. This method is, in fact, a generalization of the linear combination of atomic orbitals discussed in Chapter 7.

Let us denote the position of a unit cell by the vector $\boldsymbol{a}$ and the position of an atom in the cell by the vector $\boldsymbol{l}$. The position of the $\boldsymbol{l}$-th atom in the unit cell $\boldsymbol{a}$ will thus be defined by the vector $\boldsymbol{a} + \boldsymbol{l}$. Let $\varphi_j(\boldsymbol{r} - \boldsymbol{a} - \boldsymbol{l})$ be the wave functions of the $\boldsymbol{l}$-th atom in the unit cell $\boldsymbol{a}$. We can then readily construct linear combinations of these functions which will transform in accordance with an irreducible representation of the translation group:

$$\Phi^{(l)}_{jk}(\boldsymbol{r}) = \sum_{\boldsymbol{a}} e^{i(\boldsymbol{ka})} \varphi^{(l)}_j(\boldsymbol{r} - \boldsymbol{a} - \boldsymbol{l}) \tag{9.17}$$

In fact, we have

$$\hat{t}_a \Phi_{jk}^{(l)} = \Phi_{jk}^{(l)}(\boldsymbol{r}+\boldsymbol{a}) = \sum_{\boldsymbol{a}} e^{i\boldsymbol{k}\boldsymbol{a}} \varphi_j^{(l)}(\boldsymbol{r}+\boldsymbol{a}'-\boldsymbol{a}-\boldsymbol{l}) \qquad (9.18)$$

Let $\boldsymbol{a}'' = \boldsymbol{a} - \boldsymbol{a}'$ and let us replace summation over $\boldsymbol{a}$ by summation over $\boldsymbol{a}''$:

$$\hat{t}_{a'} \Phi_{jk}^{(l)}(\boldsymbol{r}) = \sum_{\boldsymbol{a}''} e^{i(\boldsymbol{k}(\boldsymbol{a}''+\boldsymbol{a}'))} \varphi_j^{(l)}(\boldsymbol{r}-\boldsymbol{a}''-\boldsymbol{l}) = e^{i(\boldsymbol{k}\boldsymbol{a}')} \Phi_{jk}^{(l)}(\boldsymbol{r}) \qquad (9.19)$$

The function $\Psi_{k\alpha j}(\boldsymbol{r})$ can be written in the form

$$\Psi_{k\alpha j}(\boldsymbol{r}) = \sum_{l,\, j} b_{klj} \Phi_{jk}^{(l)}(\boldsymbol{r}) \qquad (9.20)$$

where the coefficients b_{klj} are to be determined. In practice, the expansion for the function $\Psi_{k\alpha j}$ is restricted to a finite number of atomic wave functions for each cell and, consequently, a finite number of functions $\Phi_{jk}^{(l)}$ in (9.20). The roots of the corresponding secular equation give the approximate one-electron energy bands.

The solution of the secular equation can be simplified by first forming linear combinations $\tilde{\Psi}_{k\alpha j}$ from the functions $\Phi_{jk}^{(l)}$, which transform in accordance with an irreducible representation of the group of the wave vector $\boldsymbol{k}$. This can be carried out by the method given in Chapter 7. However, we must first establish how the functions transform under transformations in the group F_k. Let $g \in F_k$ and let us apply the operator $\hat{T}_g$ to the function $\Phi_{jk}^{(l)}$. The result is

$$\hat{T}_g \Phi_{jk}^{(l)} = \sum_{\boldsymbol{a}} e^{i(\boldsymbol{k}\boldsymbol{a})} \varphi_j(g^{-1}\boldsymbol{r} - \boldsymbol{a} - \boldsymbol{l})$$
$$= \sum_{\boldsymbol{a}} e^{i(\boldsymbol{k}\boldsymbol{a})} \varphi_j(g^{-1}(\boldsymbol{r} - g\boldsymbol{a} - g\boldsymbol{l})) \qquad (9.21)$$

It is clear that the vector $g\boldsymbol{a}$ is again a lattice vector:

$$g\boldsymbol{a} = \boldsymbol{a}' \qquad (9.22)$$

If, under the transformation g, the $\boldsymbol{l}$-th atom in the zero cell is transformed into an equivalent $\boldsymbol{l}'$-th atom in the cell $\boldsymbol{a}_1$, we may write

$$g\boldsymbol{l} = \boldsymbol{a}_1 + \boldsymbol{l}' \qquad (9.23)$$

We thus have

$$\hat{T}_g\Phi_{jk}^{(l)} = \sum_a e^{i(ka)}\varphi_j(g^{-1}(r - a' - a_1 - l')) \qquad (9.24)$$

Let us now suppose that the transformation rule for the atomic wave functions under the operation g is known, i.e.

$$\varphi_j^{(l)}(g^{-1}r) = \sum_i c_{ij}(g)\varphi_i^{(l)}(r) \qquad (9.25)$$

Using this relationship, we have

$$\hat{T}_g\Phi_{jk}^{(l)} = \sum_a e^{i(ka)}\sum_i c_{ij}(g)\varphi_i^{(l')}(r - a'') \qquad (9.26)$$

where

$$a'' = a' + a_1 = ga + a_1 \qquad (9.27)$$

If we replace summation over a by summation over a'', we obtain

$$\hat{T}_g\Phi_{jk}^{(l)} = \sum_{a''} e^{i(k, g^{-1}(a''-a_1))}\sum_i c_{ij}(g)\varphi_i^{(l')}(r - a'') \qquad (9.28)$$

We shall now use the fact that the orthogonal transformation g belongs to the group of the wave vector. Consequently,

$$e^{i(k, g^{-1}(a''-a_1))} = e^{i(gk, (a''-a_1))} = e^{i(k, (a''-a_1))}$$

and hence, finally,

$$\hat{T}_g\Phi_{jk}^{(l)} = e^{-i(ka)}\sum_i c_{ij}(g)\Phi_{ik}^{(l')} \qquad (9.29)$$

It follows that the functions $\Phi_{jk}^{(l)}$ transform in the same way as the atomic functions in the molecular problem under transformations in the group F_k, except that if for a given transformation g the l-th atom of the zero cell is transformed into the l-th atom of the cell a, the functions $\Phi_{ik}^{(l)}$, whose linear combination replace $\Phi_{jk}^{(l)}$, assume the additional factor $e^{-i(ka_1)}$.

If the lattice is monatomic, i.e. there is only one atom in each unit cell, the functions $\Psi_{k\alpha i}$, which transform in accordance with the irreducible representations in F_k, can be written as linear combinations $\tilde{\varphi}_{\alpha i}$ of atomic functions which transform in accordance with these irreducible representations. In point of fact, it is readily verified that, in this case, the function $\tilde{\Psi}_{kai}$ can be written in the form

$$\tilde{\Psi}_{kai}(\boldsymbol{r}) = \sum_{\boldsymbol{a}} e^{i(\boldsymbol{k}\boldsymbol{a})} \tilde{\varphi}_{ai}(\boldsymbol{r} - \boldsymbol{a}) \tag{9.30}$$

If the decomposition of the representation of F_k, which governs the transformation of the functions $\Phi_{jk}^{(l)}$, contains a particular irreducible representation only once, the functions $\tilde{\Psi}_{kai}$ corresponding to it will be the approximate one-electron wave functions for the problem. In the general case, when an irreducible representation $\Gamma^{(\alpha)}$ is encountered several times in the above representation, the wave functions $\tilde{\Psi}_{kai}$ must be determined by solving a secular equation whose order is equal to the multiplicity of the representation $\Gamma^{(\alpha)}$.

Exercise

9.1. Derive the classification of the normal vibrations of a diatomic crystal with a simple cubic lattice (Fig. 14).

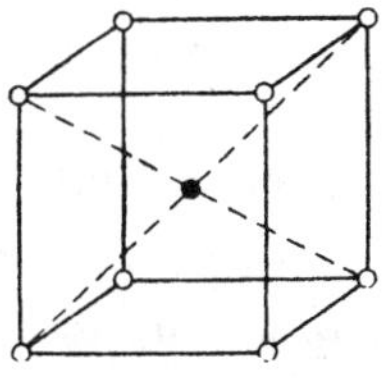

Fig. 14

Chapter 10

Continuous Groups

10.1 Continuous groups of linear transformations

We shall consider groups of linear transformations such that the elements of their matrices are analytical functions of real parameters. Let $g(\alpha_1, \alpha_2, \ldots, \alpha_r)$ be an element of such a group. The parameters $\alpha_1, \alpha_2, \ldots, \alpha_r$ are chosen so that there is a one-to-one correspondence between the neighborhood of the origin in the r-dimensional space of the parameters and the neighborhood of the unit element of the group. If

$$g(\alpha_1, \alpha_2, \ldots, \alpha_r)\, g(\alpha_1', \alpha_2', \ldots, \alpha_r') = g(\alpha_1'', \alpha_2'', \ldots, \alpha_r'') \quad (10.1)$$

then

$$\alpha_k'' = \varphi_k(\alpha_1, \alpha_2, \ldots, \alpha_r;\, \alpha_1', \alpha_2', \ldots, \alpha_r') \quad (10.2)$$

The functions φ_k which define the group multiplication are assumed to be differentiable with respect to all the arguments. Moreover, they are subject to the restriction imposed by the general group postulates. The groups of linear transformations which we shall discuss and which satisfy these requirements belong to the class of continuous Lie groups. If the parameters α_i vary within a bounded region of

r-dimensional space, the group is compact. Let us now list some of the Lie groups of linear transformations:

1. The whole linear group $GL(n)$ consists of non-singular complex matrices of order n. The elements of this group depend on $2n^2$ real parameters.

2. The unimodular group $SL(n)$ consists of all the complex matrices of order n whose determinant is equal to unity. For this group $r = 2n^2 - 2$. Its real sub-group depends on $n^2 - 1$ parameters.

3. The unitary group $U(n)$ consists of unitary matrices of order n. Since n^2 orthogonality and normalization conditions are imposed on the elements of a unitary matrix, the number of parameters which determine an arbitrary element of the group $U(n)$ is $2n^2 - n^2 = n^2$. The unitary group is a compact group because the sum of the squares of the moduli of the elements of an n-th order unitary matrix is equal to n. Transformations belonging to the unitary group preserve the quadratic form

$$x_1\bar{x}_1 + x_2\bar{x}_2 + \dots + x_n\bar{x}_n$$

4. The unitary unimodular group $SU(n)$ is a sub-group of $U(n)$. It consists of unitary matrices with determinants equal to unity. The number of its parameters is $r = n^2 - 1$.

5. The orthogonal group $O(n)$ is a real sub-group of $U(n)$. $n + \frac{n(n-1)}{2}$ orthogonality and normalization conditions are imposed on the elements of an arbitrary orthogonal matrix and, therefore, the number of parameters in this group is $\frac{n(n-1)}{2}$. The determinant of an orthogonal matrix is equal to 1 or -1.*

---

* Editor's note. The group $O^+(n)$ falls into two separate parts which cannot be transformed continuously into each other: these are the sets of real orthogonal matrices with determinant + 1 and −1 respectively. For such groups the relations (10.1) and (10.2) require generalization, as do many of the statements of this chapter.

6. The rotation group $O^+(n)$ consists of orthogonal matrices of order n with determinants equal to unity. It is clear that the number of parameters in this group is also equal to $\frac{n(n-1)}{2}$. The three-dimensional rotation group $O^+(3)$ is of particular interest in physical applications.

10.2 General properties of Lie groups

Consider a group G of linear transformations with matrix elements $g_{ik} = g_{ik}(\alpha_1, \alpha_2, \ldots, \alpha_r)$. Let us introduce the derivatives of these matrices with respect to the parameters α_l at the point $\alpha_k = 0$ $(k = 1, 2, \ldots, r)$, i.e. let us introduce the matrices I_l with elements

$$\{I_l\}_{ik} = \left(\frac{\partial g_{ik}}{\partial \alpha_l}\right)_0 \qquad (10.3)$$

The matrices I_l will be called the infinitesimal matrices of the group G.

In the neighborhood of the unit element, each element of the matrix g can be expanded into a series in powers of α_l. If we retain in these expansions only those terms which are linear in α_l, we can represent the general element of the group in the neighborhood of its unit element in the form

$$g = E + \sum_{l=1}^{r} \alpha_l I_l$$

There is a degree of arbitrariness in the selection of the group parameters. In fact, we can always transform to new parameters if we take for them any single-valued functions of $\alpha_1, \ldots, \alpha_r$:

$$\tilde{\alpha}_1 = \tilde{\alpha}_1(\alpha_1, \alpha_2, \ldots, \alpha_r)$$
$$\cdots\cdots\cdots\cdots$$
$$\tilde{\alpha}_r = \tilde{\alpha}_r(\alpha_1, \alpha_2, \ldots, \alpha_r)$$

for which $\left|\frac{\partial \tilde{\alpha}_i}{\partial \alpha_k}\right| \neq 0$. In particular, if we multiply each of the parameters α_l by a number λ_l so that

$$\tilde{\alpha}_l = \lambda_l \alpha_l$$

then for the new infinitesimal matrices corresponding to $\tilde{\alpha}_l$ we have

$$\tilde{I}_l = \frac{1}{\lambda_l} I_l \tag{10.4}$$

In the general case, we have

$$\tilde{I}_l = \sum_j I_j \left(\frac{\partial \alpha_j}{\partial \tilde{\alpha}_l} \right)_0 \tag{10.5}$$

Consider now the element $gg'g^{-1}$ which is the conjugate of g'. If g' is an infinitesimally small transformation (10.4), then to within terms linear in α_l we have

$$gg'g^{-1} = E + \sum_{l=1}^{r} gI_l g^{-1} \alpha_l \tag{10.6}$$

We see that the transformation $gg'g^{-1}$ can be characterized by the same values of the parameters α_l as the transformation g' if we take

$$\tilde{I}_l = gI_l g^{-1} \tag{10.7}$$

as the infinitesimal matrices. According to (10.2), the transformation

$$\tilde{\alpha}_k = \varphi_k (\gamma;\ \varphi(\alpha,\ \bar{\gamma}))$$

where γ and $\bar{\gamma}$ represent sets of values of the parameters corresponding to the matrices g and g^{-1}, corresponds to the transformation (10.6).

Let us now suppose that the parameters γ are also small quantities so that to within terms of the first order of small quantities in γ we have

$$g = E + \sum_k \gamma_k I_k, \quad g^{-1} = E - \sum_k \gamma_k I_k \tag{10.8}$$

Substituting this into (10.7) we obtain

$$\tilde{I}_s = I_s + \sum_k (I_k I_s - I_s I_k) \gamma_k \tag{10.9}$$

+ higher order terms.

According to (10.5), the left-hand side of this equation contains a linear combination of infinitesimal matrices and, therefore, since terms involving different powers of γ_k are independent, we may conclude that the commutator $I_k I_s - I_s I_k$ is a linear combination of infinitesimal matrices, i.e.

$$I_k I_s - I_s I_k = \sum_l c_{ksl} I_l \tag{10.10}$$

The coefficients c_{ksl} are the structure constants of the group.

Infinitesimal matrices define unambiguously a group, i.e. if we know these matrices we can determine any finite element of the group. The validity of this statement will now be proved for the special case when the element of a group of matrices $g(\alpha_1, \alpha_2, \ldots, \alpha_r)$ is also an element of a continuous one-parameter sub-group of this group. (It is shown in the theory of continuous groups that any group element is an element of a one-parameter sub-group, since it can be represented as a product of such elements (see, for example, L. Eisenhart, Continuous Groups of Transformations (Chapter 1)).) Let us consider a one-parameter sub-group of the group G. It is formed by a set of matrices g whose parameters may be regarded as differentiable functions of a parameter θ, i.e. $\alpha_i = \alpha_i(\theta)$. The parameter θ will be assumed to be chosen so that

$$g(\theta_1) g(\theta_2) = g(\theta_1 + \theta_2) \tag{10.11}$$

$$g(0) = E \tag{10.12}$$

Differentiating both sides of (10.11) with respect to θ_1 and then substituting $\theta_1 = 0$ and $\theta_2 = \theta$, we have

$$\frac{dg}{d\theta} = I_\theta g(\theta) \tag{10.13}$$

where $I_\theta = \left(\frac{dg}{d\theta}\right)_{\theta=0}$ is the infinitesimal matrix corresponding to the parameter θ. Equation (10.13) forms a set of ordinary differential equations for the elements of the matrix $g(\theta)$, which has a unique solution corresponding to the initial condition (10.12). This solution can be written in the form

$$g(\theta) = \exp I_\theta \theta \tag{10.14}$$

where

$$\exp I_\theta \theta = E + I_\theta \theta + \frac{1}{2}(I_\theta \theta)^2 + \frac{1}{3!}(I_\theta \theta)^3 + \dots \tag{10.15}$$

Using this expansion, we may write

$$I_\theta = \sum_j I_j \left(\frac{d\alpha_j}{d\theta}\right)_{\theta=0}$$

and hence

$$g(\theta) = \exp I_\theta \theta = \exp \sum_{j=1}^{r} I_j \left(\frac{d\alpha_j}{d\theta}\right)_{\theta=0} \theta \tag{10.14a}$$

Equations (10.10) and (10.14) refer to the elements of a group of linear transformations. It is clear that they are also valid for any group of matrices $D(g)$ forming a representation of the group G. The infinitesimal matrices of the representation D are defined by

$$A_s = \frac{\partial}{\partial \alpha_s} D(\alpha_1, \alpha_2, \dots, \alpha_r)\Big|_{\alpha=0} \tag{10.16}$$

The same commutation relations are valid for the infinitesimal matrices of the representation as for the infinitesimal matrices of the group. This is so because the structural constants of a group are uniquely determined by the group multiplication rule (10.2) which is the same for the group and for all its representations. Some of the properties of the representations of finite groups considered in the preceding chapters (for example, Schur's lemmas) did not depend on the fact that the groups were finite, and are therefore also valid for infinite groups. However, the proof of the unitarity of representations, the orthogonality properties of the matrix elements of irreducible representations and all the consequences which follow from this are based on the possibility of summing over a group which we have defined only for finite groups. For continuous groups, summation over a group must be replaced by integration with respect to the group parameters. It turns out that integration over a group can be defined only for compact groups and, therefore, the above properties of finite groups can be generalized only to compact groups.

10.3 Infinitesimal transformations and conservation laws

We showed in Chapter 5 that the invariance of a physical system under transformations in its symmetry group G can be written in the form

$$\hat{H}\hat{T}_g = \hat{T}_g\hat{H} \tag{10.17}$$

or in the matrix form

$$HD(g) = D(g)H \tag{10.17a}$$

where $\hat{H}$ is the Hamiltonian operator for the system, $\hat{T}_g$ is an element of the operator group isomorphic to G and $D(g)$ is the representation matrix of G.

If the transformation group G is continuous, the necessary and sufficient condition for (10.17) to be satisfied is

$$HA_l = A_lH \tag{10.18}$$

where A_l are infinitesimal representation matrices of G. The necessity of this condition is obvious: if for all the elements of the group G the condition (10.17) is satisfied, then by definition of the infinitesimal matrices A_l the condition (10.18) is also satisfied. It is readily verified that (10.18) is also sufficient. We have shown the validity of (10.14) and, therefore, if (10.18) is satisfied, i.e. H and A_l commute, then H will also commute with all the terms in the series corresponding to an arbitrary element of G.

Instead of the infinitesimal matrices A_j let us introduce $B_j = -iA_j$. We shall show that if the representation is unitary, then the matrices B_j are Hermitian. In fact, let us write down the condition for the unitarity of the representation $D(g)$ to within terms which are linear in the group parameters. We have

$$E = D^+(g)\,D(g) = \Big(E - i\sum_j B_j^+\alpha_j\Big)\Big(E + i\sum_j B_j\alpha_j\Big)$$

and hence

$$\sum_j \alpha_j\,(B_j^+ - B_j) = 0$$

or, since the parameters α_j are independent,

$$B_j^+ = B_j \tag{10.19}$$

The Hermitian matrices B_j also commute with H, i.e.

$$HB_j = B_jH \tag{10.18a}$$

Hermitian matrices or operators are associated with physical quantities in quantum mechanics, and the commutation relation (10.18a) then indicates that the corresponding physical quantity is a constant of motion. The conservation laws in quantum mechanics can then be regarded as a consequence of the symmetry properties of the Hamiltonian under a continuous group of transformations.

10.4 The two-dimensional rotation group O^+ (2)

The elements of this group are rotations in three-dimensional space about a particular axis or, what amounts to the same thing, rotations in the plane perpendicular to this axis about the origin. If we take the axis of rotation as the z-axis, the elements of the group $O^+(2)$ will be the transformations

$$x' = x\cos\varphi - y\sin\varphi, \quad y' = x\sin\varphi + y\cos\varphi \tag{10.20}$$

where φ is the angle of rotation which lies in the range $0 \leqslant \varphi < 2\pi$. The group $O^+(2)$ is thus a one-parameter compact group. Since it is Abelian and compact, all its first-order irreducible representations are unitary. We can therefore conclude that a number $\chi(\varphi)$ of modulus 1 can be associated in an irreducible representation with each group element. Moreover,

$$\chi(\varphi_1)\,\chi(\varphi_2) = \chi(\varphi_1 + \varphi_2), \quad \chi(0) = \chi(2\pi) \tag{10.21}$$

and hence it is readily concluded that

$$\chi(\varphi) = \exp im\varphi, \text{ where } m = 0, \pm 1, \pm 2, \ldots \qquad (10.22)$$

The group $O^+(2)$ can be regarded as the limiting case of the group C_n as $n \to \infty$. Therefore, it can occasionally be indicated by C_∞. The group $C_{\infty v} = C_\infty \times \sigma_v$ (the symmetry group of a diatomic molecule with different nuclei) and the group $D_{\infty h} = C_{\infty v} \times i$ (the symmetry group of a diatomic molecule with identical nuclei) are of particular physical interest. The stationary states and the energy levels of such molecules are classified in accordance with the irreducible representations of such groups. The characters of the irreducible representations of these groups can be obtained in the same way as the characters of the irreducible representations of the groups C_{nv} and C_{nh} (Chapter 6).

10.5 The three-dimensional rotation group $O^+(3)$

We know that this group is a three-parameter group. Let us take as these parameters the three components $\alpha_1, \alpha_2, \alpha_3$ of a vector $\boldsymbol{\alpha}$ lying along the axis of rotation and having a length equal to the angle of rotation. The direction of rotation will be defined by the right-hand rule. It is clear that we can always suppose that $|\boldsymbol{\alpha}| \leqslant \pi$ and, therefore, the range of the parameters α_i is a sphere of radius π, i.e. the group $O^+(3)$ is compact. We note that different internal points of the sphere correspond to different rotations, while any two points on the surface of the sphere lying at the opposite ends of a diameter correspond to the same rotation (through an angle π).

Let us determine the infinitesimal matrices I_i of the rotation group. It is clear that when the matrix $g(\alpha_1, \alpha_2, \alpha_3)$, which is an element of the group $O^+(3)$, is differentiated with respect to one of the parameters, we can set the remaining parameters equal to zero. Therefore, to calculate I_1, let us consider the matrix corresponding to the vector

$a(\alpha_1, 0, 0)$, i.e. rotation through an angle α_1 about the x-axis. This will be the matrix

$$g(\alpha_1, 0, 0) = \begin{pmatrix} 1 & 0 & 0 \\ 0 & \cos\alpha_1 & -\sin\alpha_1 \\ 0 & \sin\alpha_1 & \cos\alpha_1 \end{pmatrix}$$

and, consequently,

$$I_1 = \frac{\partial g(\alpha_1, 0, 0)}{\partial \alpha_1}\bigg|_{\alpha_1=0} = \begin{pmatrix} 0 & 0 & 0 \\ 0 & 0 & -1 \\ 0 & 1 & 0 \end{pmatrix}$$

Similarly, we find that

$$I_2 = \begin{pmatrix} 0 & 0 & 1 \\ 0 & 0 & 0 \\ -1 & 0 & 0 \end{pmatrix}, \quad I_3 = \begin{pmatrix} 0 & -1 & 0 \\ 1 & 0 & 0 \\ 0 & 0 & 0 \end{pmatrix}$$

It is readily verified that these matrices satisfy the following commutation relations:

$$\left.\begin{aligned} I_1 I_2 - I_2 I_1 &= I_3 \\ I_2 I_3 - I_3 I_2 &= I_1 \\ I_3 I_1 - I_1 I_3 &= I_2 \end{aligned}\right\} \tag{10.23}$$

We shall show that the increment of an arbitrary vector due to rotation through an infinitesimal angle about a given axis can be expressed in terms of the matrices I_i. Thus, consider the vector

$$\boldsymbol{r}' = g(\alpha_1, \alpha_2, \alpha_3)\,\boldsymbol{r} = \boldsymbol{r} + \delta\boldsymbol{r}$$

where $g(\alpha_1, \alpha_2, \alpha_3)$ is an arbitrary element of the group $O^+(3)$. For small values of the parameters we can use (10.4) and write

$$\boldsymbol{r}' = \{E + \alpha_1 I_1 + \alpha_2 I_2 + \alpha_3 I_3\}\,\boldsymbol{r} \tag{10.24}$$

In the case of rotation about one of the coordinate axes – for example, the x-axis – we have to within terms of the first order in the small quantities α_1,

$$\delta\boldsymbol{r} = \alpha_1 I_1 \boldsymbol{r} = \begin{pmatrix} 0 \\ -\alpha_1 x_3 \\ \alpha_1 x_2 \end{pmatrix} \tag{10.25}$$

and, consequently, the change in the vector $\boldsymbol{r}$ resulting from rotation through a small angle about an axis corresponding to the parameter α_l is determined by the matrix $\alpha_l I_l$.

We shall now show how any matrix corresponding to a rotation with given values of the parameters $\alpha_1, \alpha_2, \alpha_3$ can be expressed in terms of infinitesimal matrices. Consider the one-parameter group of rotations about the axis lying along the vector $\boldsymbol{\alpha}(\alpha_1, \alpha_2, \alpha_3)$. The required matrix corresponds to rotation about this axis through an angle $\alpha = \sqrt{\alpha_1^2 + \alpha_2^2 + \alpha_3^2}$, where

$$\alpha_1 = \alpha \cos(\widehat{Ox, \alpha}), \quad \alpha_2 = \alpha \cos(\widehat{Oy, \alpha})$$
$$\alpha_3 = \alpha \cos(\widehat{Oz, \alpha}) \tag{10.26}$$

Using (10.14a), we have

$$g(\alpha_1, \alpha_2, \alpha_3) = \exp I_\alpha \alpha$$
$$= \exp\{I_1 \cos(\widehat{Ox, \alpha}) + I_2 \cos(\widehat{Oy, \alpha}) + I_3 \cos(\widehat{Oz, \alpha})\}\alpha$$
$$= \exp(I_1\alpha_1 + I_2\alpha_2 + I_3\alpha_3) \tag{10.27}$$

Exercises

10. 1. Find the matrices of the irreducible representations of the group $C_{\infty v}$.

10. 2. Find the matrices of the irreducible representations of the group $D_{\infty h}$.

Chapter 11

Irreducible Representations of the Three-Dimensional Rotation Group

11.1 Infinitesimal representation matrices of the group $O^+(3)$

In the last chapter we found the infinitesimal matrices of the group $O^+(3)$ and showed that they satisfied commutation relations (10.23) which may be represented by the vector product symbol

$$[I, I] = I \tag{11.1}$$

We note that infinitesimal matrices of group representations should satisfy the same commutation relations, and we shall denote these matrices by the letters A_1, A_2, A_3. From the results obtained in the preceding chapter it follows that if we know the infinitesimal matrices A_i of a representation, then the representation matrix corresponding to an arbitrary rotation with parameters α_1, α_2, α_3 can be written in the form

$$D(\alpha_1, \alpha_2, \alpha_3) = \exp[\alpha_1 A_1 + \alpha_2 A_2 + \alpha_3 A_3] \tag{11.2}$$

It follows that the search for the representations of the group $O^+(3)$ reduces to finding the matrices A_1, A_2, A_3 satisfying the commutation relations (11.1).

Let us now consider certain properties of the matrices A_i. The rotation group is compact and, consequently, any of its representations is equivalent to the unitary representation. We may therefore restrict our attention to unitary representations. Let $D(\alpha_1, \alpha_2, \alpha_3)$ be a unitary representation, i.e.

$$D^+(\alpha_1, \alpha_2, \alpha_3)\, D(\alpha_1, \alpha_2, \alpha_3) = E \tag{11.3}$$

If we retain only linear terms in α_i we obtain

$$(E + \alpha_1 A_1^+ + \alpha_2 A_2^+ + \alpha_3 A_3^+)(E + \alpha_1 A_1 + \alpha_2 A_2 + \alpha_3 A_3) = E$$

and hence

$$A_i^+ = -A_i \tag{11.4}$$

The matrices A_i are therefore anti-Hermitian.

It will be convenient to consider the following linear combinations of the matrices A_i instead of the matrices themselves:

$$H_+ = iA_1 - A_2, \quad H_- = iA_1 + A_2, \quad H_3 = iA_3 \tag{11.5}$$

It is readily seen that the matrix H_3 is Hermitian, and the matrices H_+ and H_- are Hermitian conjugates. Using (11.1) we can show that these matrices satisfy the following commutation relations:

$$\left.\begin{aligned} H_+H_- - H_-H_+ &= 2H_3 \\ H_-H_3 - H_3H_- &= H_- \\ H_3H_+ - H_+H_3 &= H_+ \end{aligned}\right\} \tag{11.6}$$

Let us now establish the properties of the eigenvectors of the matrix H_3. We will show first that if v_λ is an eigenvector of the matrix H_3 which corresponds to an eigenvalue λ, then H_+v_λ will be an eigenvector of H_3 corresponding to the eigenvalue $\lambda + 1$. In fact, let

$$H_3 v_\lambda = \lambda v_\lambda \tag{11.7}$$

in which case

$$H_3 H_+ v_\lambda = (H_+ H_3 + H_+) v_\lambda = (\lambda + 1) H_3 v_\lambda \tag{11.8}$$

Similarly, we can show that H_-v_λ is an eigenvector of the matrix H_3 corresponding to the eigenvalue $\lambda - 1$. We thus

have

$$H_3H_-v_\lambda = (H_-H_3 - H_-)v_\lambda = (\lambda - 1)H_-v_\lambda \tag{11.9}$$

Since the matrix H_3 is Hermitian, all its eigenvalues are real and the eigenvectors corresponding to different eigenvalues are orthogonal.

We shall now show that all the eigenvalues of the matrix H_3 are integers or half-integers differing by unity, and their maximum and minimum eigenvalues are j and $-j$, respectively.

We shall suppose that all the eigenvectors v_λ of the matrix H_3 are normalized to unity. In that case, in view of (11.8) and (11.9), we can write

$$H_+v_\lambda = \beta_\lambda v_{\lambda+1} \tag{11.10}$$

$$H_-v_\lambda = \alpha_\lambda v_{\lambda-1} \tag{11.11}$$

where β_λ and α_λ are numbers which we shall find from the normalization condition for v_λ. Since the matrices H_+ and H_- are Hermitian conjugates, we have

$$(H_+v_k, v_{k+1}) = \beta_k(v_{k+1}, v_{k+1}) = \beta_k$$
$$= (v_k, H_-v_{k+1}) = \alpha_{k+1}(v_k, v_k) = \alpha_{k+1}$$

and, therefore,

$$\beta_k = \alpha_{k+1} \tag{11.12}$$

To establish explicit expressions for these coefficients, let us find the recurrence relation for the coefficients β_k. When $k < j$, we can write

$$H_+v_{k-1} = \frac{1}{\alpha_k}H_+H_-v_k = \frac{1}{\alpha_k}(H_-H_+ + 2H_3)v_k$$
$$= \frac{1}{\alpha_k}(\alpha_{k+1}\beta_k + 2k)v_k = \beta_{k-1}v_k$$

and hence, using (11.12), we have

$$\beta_k^2 + 2k = \beta_{k-1}^2 \tag{11.13}$$

If, on the other hand, $k = j$, then $H_+v_k = 0$, and, consequently,

$$\beta_{j-1}^2 = 2j \tag{11.14}$$

Using (11.13) and (11.14), we find by induction that

$$\beta_k^2 = j(j+1) - k(k+1) \tag{11.15}$$

and hence, using (11.12), we find that

$$\alpha_k^2 = j(j+1) - k(k-1) \tag{11.16}$$

It follows that the effect of the matrices H_+, H_-, H_3 on the vectors v_k is as follows:

$$\left.\begin{aligned} H_3 v_k &= k v_k \\ H_- v_k &= \sqrt{j(j+1) - k(k-1)}\, v_{k-1} \\ H_+ v_k &= \sqrt{j(j+1) - k(k+1)}\, v_{k+1} \end{aligned}\right\} \tag{11.17}$$

By applying successively different powers of the matrix H_- to the eigenvector v_j, we obtain the set of eigenvectors

$$v_j,\ v_{j-1},\ v_{j-2},\ \ldots \tag{11.18}$$

In view of (11.17), the last vector in this set will be the vector v_{-j}, since $H_- v_{-j} = 0$. Hence, it follows that the smallest eigenvalue of the matrix H_3 is $-j$. The number of all the eigenvectors is equal to $2j+1$ and, therefore, j can be either an integer or a half-integer.

11.2 Irreducible representations of the group $O^+(3)$

The orthonormal eigenvectors

$$v_j,\ v_{j-1},\ v_{j-2},\ \ldots,\ v_{-j}$$

of the matrix H_3 form the basis of the space R_{2j+1}, which we shall call the canonical basis. From (11.17) it follows that R_{2j+1} is invariant under transformations with matrices H_+, H_-, H_3 and, consequently, a certain representation of the group $O^+(3)$ is realized in it. We shall show that this representation is irreducible. To establish this, we shall prove that the space R_{2j+1} has no invariant sub-spaces with respect to transformations with matrices $D(\alpha_1, \alpha_2, \alpha_3)$ of this representation. Let us assume the opposite, i.e. that there exists a sub-space $R' \subset R_{2j+1}$ which is invariant under all the transformations $D(\alpha_1, \alpha_2, \alpha_3)$. If this is so, the sub-space should be also invariant under infinitesimal transformations A_i (or H_+, H_-, H_3). Let $\boldsymbol{h} \in R'$ be the eigenvector of the matrix H_3 corresponding to the maximum eigenvalue. Since $R' \subset R_{2j+1}$, the vector $\boldsymbol{h}$ can be represented by a linear combination of the vectors $\boldsymbol{v}_k$:

$$\boldsymbol{h} = \sum_{k=-j}^{j} c_k \boldsymbol{v}_k \tag{11.19}$$

It is clear that

$$H_+ \boldsymbol{h} = 0 \tag{11.20}$$

Consequently, we have

$$H_+ \boldsymbol{h} = \sum_{k=-j}^{j} c_k H_+ \boldsymbol{v}_k = \sum_{k=-j}^{j} c_k \beta_k \boldsymbol{v}_{k+1} = 0 \tag{11.21}$$

and since the vectors $\boldsymbol{v}_k$ are linearly independent,

$$c_{-j} = c_{-j+1} = \ldots = c_{j-1} = 0 \tag{11.22}$$

We thus have

$$\boldsymbol{h} = c_j \boldsymbol{v}_j \tag{11.23}$$

and, consequently, $\boldsymbol{v}_j \in R'$. However, in that case, all the vectors $\boldsymbol{v}_{j-1}, \boldsymbol{v}_{j-2}, \ldots, \boldsymbol{v}_{-j}$ should also belong to R', i.e. R' coincides with the space R_{2j+1}. We have thus proved that the representation realized in the space R_{2j+1} is irreducible.

We see that an irreducible representation of the group $O^+(3)$ is determined by the maximum eigenvalue j of the matrix H_3. The number j is called the weight of the irreducible representation. An irreducible representation of weight j will be denoted by $D^{(j)}$. It is clear that the order of the representation $D^{(j)}$ is $2j+1$.

We shall show that all the basis vectors of an irreducible representation are eigenvectors of the matrix

$$A^2 = A_1^2 + A_2^2 + A_3^2 \tag{11.24}$$

corresponding to the eigenvalue $-j(j+1)$. If we express the matrix A^2 in terms of the matrices H_+, H_-, H_3, we shall

have

$$A^2 v_k = -(H_+ H_- - H_3 + H_3^2) v_k = (\beta_k^2 - k + k^2) v_k$$

and hence

$$A^2 v_k = -j(j+1) v_k \qquad (k = -j, -j+1, \ldots, j) \tag{11.25}$$

Let us now determine the form of the infinitesimal matrices A_l of an irreducible representation $D^{(j)}$ in the canonical basis. From (11.5) and (11.17) we have

$$\left.\begin{aligned} A_1 v_k &= -\frac{i}{2}\{\alpha_k v_{k-1} + \alpha_{k+1} v_{k+1}\} \\ A_2 v_k &= \frac{1}{2}\{\alpha_k v_{k-1} - \alpha_{k+1} v_{k+1}\} \\ A_3 v_k &= -ikv_k \end{aligned}\right\} \tag{11.26}$$

Hence

$$\left.\begin{aligned} \{A_1\}_{kl} &= (v_l, A_1 v_k) = -\frac{i}{2}\{\alpha_k \delta_{l,k-1} + \alpha_{k+1}\delta_{l,k+1}\} \\ \{A_2\}_{kl} &= (v_l, A_2 v_k) = \frac{1}{2}\{\alpha_k \delta_{l,k-1} - \alpha_{k+1}\delta_{l,k+1}\} \\ \{A_3\}_{kl} &= (v_l, A_3 v_k) = -ik\delta_{lk}, \quad k, l = -j, -j+1, \ldots, j \end{aligned}\right\} \tag{11.27}$$

and, consequently, the matrices A_l have the form

$$A_1 = \left\{\begin{array}{ccccccc} 0 & -\frac{i}{2}\alpha_{-j+1} & 0 & \ldots & 0 & 0 & 0 \\ -\frac{i}{2}\alpha_{-j+1} & 0 & -\frac{i}{2}\alpha_{-j+2} & \ldots & 0 & 0 & 0 \\ 0 & -\frac{i}{2}\alpha_{-j+2} & 0 & \ldots & 0 & 0 & 0 \\ \cdots & \cdots & \cdots & \cdots & \cdots & \cdots & \cdots \\ 0 & 0 & 0 & \ldots & -\frac{i}{2}\alpha_{j-1} & 0 & -\frac{i}{2}\alpha_j \\ 0 & 0 & 0 & \ldots & 0 & -\frac{i}{2}\alpha_j & 0 \end{array}\right\}$$

$$A_2 = \left\{\begin{array}{ccccccc} 0 & -\frac{1}{2}\alpha_{-j+1} & 0 & \ldots & 0 & 0 & 0 \\ \frac{1}{2}\alpha_{-j+1} & 0 & -\frac{1}{2}\alpha_{-j+2} & \ldots & 0 & 0 & 0 \\ 0 & \frac{1}{2}\alpha_{-j+2} & 0 & \ldots & 0 & 0 & 0 \\ \cdots & \cdots & \cdots & \cdots & \cdots & \cdots & \cdots \\ 0 & 0 & 0 & \ldots & \frac{1}{2}\alpha_{j-1} & 0 & -\frac{1}{2}\alpha_j \\ 0 & 0 & 0 & \ldots & 0 & \frac{1}{2}\alpha_j & 0 \end{array}\right\} \tag{11.28}$$

$$A_3 = \begin{Vmatrix} ij & 0 & \ldots & 0 \\ 0 & i(j-1) & \ldots & 0 \\ \vdots & & \ddots & \vdots \\ 0 & 0 & \ldots & -ij \end{Vmatrix} \tag{11.28}$$

To determine the matrices $D^{(j)}(\alpha_1, \alpha_2, \alpha_3)$ we can use the formula (11.2)

$$D^{(j)}(\alpha_1, \alpha_2, \alpha_3) = e^{\alpha_1 A_1 + \alpha_2 A_2 + \alpha_3 A_3} \tag{11.29}$$

Let us determine, as an example, the form of the matrix $D^{(j)}(0, 0, \alpha_3)$ in the canonical basis. According to (11.29) we have

$$D^{(j)}(0, 0, \alpha_3) = e^{\alpha_3 A_3} \tag{11.30}$$

Since in the canonical basis the matrix A_3 is diagonal (see (11.28)), it follows that all its powers are also diagonal and, consequently, by writing (11.30) in the form of a series, and then summing the series corresponding to identical matrix elements, we can write

$$D^{(j)}(0, 0, \alpha_3) = \begin{Vmatrix} e^{ij\alpha_3} & 0 & \ldots & 0 \\ 0 & e^{i(j-1)\alpha_3} & \ldots & 0 \\ \cdot & \cdot & \cdot & \cdot \\ 0 & 0 & \ldots & e^{-ij\alpha_3} \end{Vmatrix} \tag{11.31}$$

This result enables us to determine the characters of the irreducible representation $D^{(j)}$. In fact, rotations about different axes through the same angle are included in a single class of the rotation group. The class is determined only by the angle of rotation, and rotation through an angle φ about the z-axis can be taken as representing the class. From (11.31) it follows that the character of the class of rotations through an angle φ is

$$\chi^{(j)}(\varphi) = \sum_{l=-j}^{j} e^{il\varphi} = \frac{\sin\left(j + \frac{1}{2}\right)\varphi}{\sin\frac{\varphi}{2}} \tag{11.32}$$

11.3 Two-valued representations

The matrix (11.31) corresponds to rotation through an angle α_3 about the z-axis. The unit matrix E_{2j+1} corresponds to a unit element of the group $O^+(3)$ in the representation by the matrices $D^{(j)}$. Rotation through 2π about the z-axis is also a unit element of the group $O^+(3)$. For an integral j we have $D^{(j)}(0, 0, 2\pi) = E_{2j+1}$ and for half-integral j the result is $D^{(j)}(0, 0, 2\pi) = -E_{2j+1}$. Therefore, for half-integral j the two matrices E and $-E$ correspond to the unit element of the group $O^+(3)$ and, consequently, to each element of this group there correspond two matrices $D^{(j)}$ and $-D^{(j)}$, whose elements differ in sign. In the case of half-integral j, the matrices $D^{(j)}$ are said to give a two-valued representation of the group $O^+(3)$. Two-valued representations play a very important role in physical applications. They are used, as we shall see in due course, for the description of particles with half-integral spins.

Let us now establish the explicit form of the matrices $D^{(1/2)}$. We know that any rotation can be achieved as a result of the following three successive rotations: rotation about the z-axis through an angle φ_1, rotation about the x-axis through an angle θ and rotation about the z-axis through an angle φ_2 (Fig.15). The matrix $D^{(1/2)}[\varphi_1, \theta, \varphi_2]$ can therefore be written in the form

$$D^{(1/2)}[\varphi_1, \theta, \varphi_2] = D^{(1/2)}(0, 0, \varphi_2)\, D^{(1/2)}(\theta, 0, 0)\, D^{(1/2)}(0, 0, \varphi_1) \quad (11.33)$$

where the parameters $\varphi_1, \theta, \varphi_2$ are the Euler angles. According to (11.31) the matrix $D^{(1/2)}(0, 0, \varphi)$ which corresponds to rotation about the z-axis through an angle φ is of the form

$$D^{(1/2)}(0, 0, \varphi) = \begin{pmatrix} e^{i\frac{\varphi}{2}} & 0 \\ 0 & e^{-i\frac{\varphi}{2}} \end{pmatrix} \quad (11.34)$$

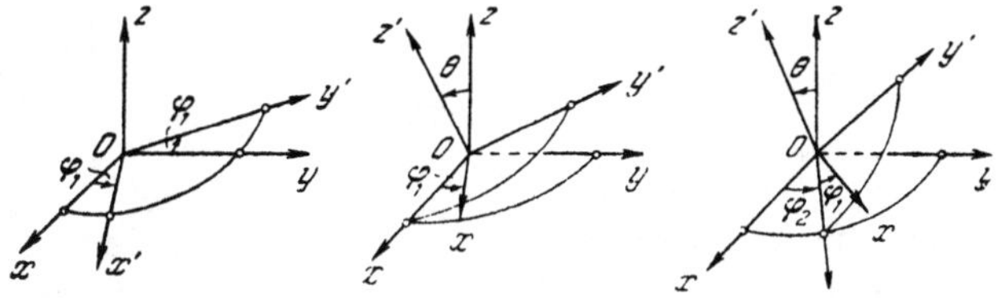

Fig.15

The matrix $D^{(1/2)}(\theta, 0, 0)$, corresponding to rotation through an angle x about the θ-axis, can be found by solving a set of differential equations analogous to (10.13), namely,

$$\frac{dD}{d\theta} = A_1 D \tag{11.35}$$

subject to the initial condition $D(0) = E$. Since the infinitesimal matrix A_1 for the representation $D^{(1/2)}$ is according to (11 .28) given by

$$A_1 = -\frac{i}{2}\begin{pmatrix} 0 & 1 \\ 1 & 0 \end{pmatrix}$$

the system given by (11.35) can be written in the form

$$\left.\begin{aligned} \frac{dD_{11}}{d\theta} &= -\frac{i}{2}D_{21}, & \frac{dD_{12}}{d\theta} &= -\frac{i}{2}D_{22} \\ \frac{dD_{21}}{d\theta} &= -\frac{i}{2}D_{11}, & \frac{dD_{22}}{d\theta} &= -\frac{i}{2}D_{12} \end{aligned}\right\} \tag{11.36}$$

and the initial conditions can be written

$$D_{11}(0) = D_{22}(0) = 1, \quad D_{12}(0) = D_{21}(0) = 0$$

It is readily verified that the solution of this system, which satisfies the initial conditions, gives

$$D^{(1/2)}(\theta, 0, 0) = \begin{pmatrix} \cos\frac{\theta}{2} & i\sin\frac{\theta}{2} \\ i\sin\frac{\theta}{2} & \cos\frac{\theta}{2} \end{pmatrix} \tag{11.37}$$

Substituting (11.34) and (11.37) into (11.33), we obtain the matrix $D^{(1/2)}(\varphi_1, \theta, \varphi_2)$ for an arbitrary rotation:

$$D^{(1/2)}[\varphi_1, \theta, \varphi_2] = \begin{pmatrix} \cos\frac{\theta}{2} e^{i\frac{\varphi_1+\varphi_2}{2}} & i\sin\frac{\theta}{2} e^{-i\frac{\varphi_2-\varphi_1}{2}} \\ i\sin\frac{\theta}{2} e^{i\frac{\varphi_2-\varphi_1}{2}} & \cos\frac{\theta}{2} e^{-i\frac{\varphi_1+\varphi_2}{2}} \end{pmatrix} \quad (11.38)$$

The matrix (11.38) is unitary and, moreover, its determinant is equal to unity. It is readily verified that any unitary unimodular matrix of order two can be represented in this form. Hence, it may be concluded that to any unitary matrix of order two and unit determinant there corresponds a rotation in three-dimensional space. Conversely, to any rotation of three-dimensional space there correspond two matrices which differ only in sign. Therefore, the rotation group O (3) is homomorphic to the group SU (2) of unitary unimodular matrices of order 2.

11.4 Decomposition of any representation of the group O^+ (3) into irreducible representations

Let R_n be the basis space of a representation by matrices D of order n. We shall define the matrices H_+, H_- and H_3 for this representation and find the greatest eigenvalue j' of the matrix H_3. Using the matrix H_-, let us form the chain of eigenvectors $\boldsymbol{v}_{j'}, \boldsymbol{v}_{j'-1}, \dots, \boldsymbol{v}_{-j'}$ of the matrix H_3, corresponding to eigenvalues differing by unity. We shall take these vectors as the first $2j'+1$ basis vectors in the new basis of the space R_n; the remaining unit vectors of this basis, which define the sub-space R', will be taken arbitrarily for the time being. The space R_n can be regarded as the direct sum $R_{2j'+1}$ and R':

$$R_n = R_{2j'+1} \oplus R'$$

In the new basis, the matrices D assume the form

$$U^{-1} DU = \begin{pmatrix} D^{(j')} & 0 \\ 0 & D' \end{pmatrix}$$

Let us now find the maximum eigenvalue of the matrix H_3 and the corresponding eigenvector in the space R'. We shall denote these quantities by j'' and $\boldsymbol{v}_{j''}$, respectively. If j' is degenerate, then $j''=j'$. If, on the other hand, j' is a simple eigenvalue, then $j''<j'$. Consider the chain of eigenvectors of H_3:

$$\boldsymbol{v}_{j''},\ \boldsymbol{v}_{j''-1},\ \ldots,\ \boldsymbol{v}_{-j''}$$

Assuming that the first $2j'+1$ unit vectors of the new basis are fixed, let us take the vectors of this chain as the next $2j''+1$ unit vectors of the new basis in R_n. The remaining $n-(2j'+1)-(2j''+1)$ unit vectors taken arbitrarily define the sub-space R''. We now have

$$R_n = R_{2j'+1} \oplus R_{2j''+1} \oplus R''$$

and the matrices D assume the form

$$U^{-1}DU = \begin{vmatrix} D^{(j')} & 0 & 0 \\ 0 & D^{(j'')} & 0 \\ 0 & 0 & D'' \end{vmatrix}$$

Continuing this process, we shall exhaust the space R_n after a finite number of steps. In the final basis, the matrices D will be quasi-diagonal and will consist of the matrices $D^{(j)}$ of the irreducible representations which we have found.

11.5 Irreducible representations of the $O(3)$ orthogonal group

The complete orthogonal group $O(3)$, which is the group of orthogonal transformations in three-dimensional space, is the direct product of the rotation group $O(3)$ and the inversion group:

$$O(3) = O^{+}(3) \times \boldsymbol{i} \tag{11.39}$$

The group $\boldsymbol{i}$ has two first-order irreducible representations: the identity and the alternating representations. The group $O(3)$ has, therefore, two irreducible representations of order

$2l+1$ for each integral l. We shall denote these representations by $D_g^{(l)}$ and $D_u^{(l)}$. The matrices corresponding to rotations in these representations coincide, and the elements of matrices corresponding to rotations accompanied by reflections differ in sign. Therefore, the number of odd-order representations of the group $O(3)$ is twice as high as for the group $O^+(3)$.

The situation is different in the case of two-valued representations of the group $O^+(3)$ in which to each element $g[\varphi_1, \theta, \varphi_2]$ of this group there correspond two matrices whose elements differ in sign. It is clear that the representations of the group $O(3)$ obtained by multiplication of the two-valued representation by the identity representation of the group, and by multiplication by the alternating representation, will be identical: in each of them two matrices differing in sign will correspond to both the element g and the element ig. The group $O(3)$, like the group $O^+(3)$, has, therefore, one two-valued representation $D^{(j)}$ for each $j = \frac{2n+1}{2}$.

Chapter 12

The Properties of Irreducible Representations of the Rotation Group

12.1 Spherical harmonics as a basis for an irreducible representation

So far, we have investigated representations of the group $O^+(3)$ without specifying the space in which they are realized. We shall now consider the special case when the representation space is the space of the differentiable functions $f(\boldsymbol{n}) = f(\theta, \varphi)$ defined on the surface of a sphere of unit radius, where $\boldsymbol{n}$ is a unit vector defined by the polar angles θ and φ. As a result of a rotation g, the vector $\boldsymbol{n}$ becomes transformed to the vector $g\boldsymbol{n}$. The corresponding transformation of the functions $f(\boldsymbol{n})$ is determined by the unitary operator $\hat{T}_g$:

$$\hat{T}_g f(\boldsymbol{n}) = f(g^{-1}\boldsymbol{n}) \tag{12.1}$$

We know that the operators $\hat{T}_g$ should form a representation of the group $O^+(3)$ (see Section 3.3). Let us find the infinitesimal operators $\hat{A}_i$ of this representation. They will define the increment of the function $f(\boldsymbol{n})$ which is linear in the rotation parameters α_i. For rotation about the z-axis through an angle α we have

$$\hat{T}_g f(\theta, \varphi) = f(\theta, \varphi - \alpha) = f(\theta, \varphi) - \alpha \frac{\partial f(\theta, \varphi)}{\partial \varphi} + \ldots \quad (12.2)$$

and, consequently,

$$\hat{A}_3 f(\theta, \varphi) = -\frac{\partial f(\theta, \varphi)}{\partial \varphi} \quad (12.3)$$

For an arbitrary rotation defined by the parameters α_1, α_2, α_3, we have

$$T_g f(\theta, \varphi) = f(\theta', \varphi')$$
$$= f(\theta, \varphi) + \sum_{i=1}^{3} \left\{ \frac{\partial f}{\partial \theta} \left(\frac{\partial \theta'}{\partial \alpha_i} \right)_0 + \frac{\partial f}{\partial \varphi} \left(\frac{\partial \varphi'}{\partial \alpha_i} \right)_0 \right\} \alpha_i + \ldots \quad (12.4)$$

and hence

$$\hat{A}_i = a_i(\theta, \varphi) \frac{\partial}{\partial \theta} + b_i(\theta, \varphi) \frac{\partial}{\partial \varphi} \quad (12.5)$$

where

$$a_i(\theta, \varphi) = \left(\frac{\partial \theta'}{\partial \alpha_i} \right)_0, \quad b_i(\theta, \varphi) = \left(\frac{\partial \varphi'}{\partial \alpha_i} \right)_0 \quad (12.6)$$

Consider a rotation about the x-axis:

$$n_1' = n_1, \quad n_2' = n_2 \cos\alpha + n_3 \sin\alpha, \quad n_3' = -n_2 \sin\alpha + n_3 \cos\alpha \quad (12.7)$$

Hence, we have

$$\left.\frac{dn_1'}{d\alpha}\right|_{\alpha=0} = 0, \quad \left.\frac{dn_2'}{d\alpha}\right|_{\alpha=0} = n_3, \quad \left.\frac{dn_3'}{d\alpha}\right|_{\alpha=0} = -n_2 \quad (12.8)$$

and since

$$n_1 = \sin\theta\cos\varphi, \quad n_2 = \sin\theta\sin\varphi, \quad n_3 = \cos\theta$$

we have from (12.8)

$$\left.\begin{aligned} \cos\theta\cos\varphi \left(\frac{d\theta'}{d\alpha} \right)_0 - \sin\theta\sin\varphi \left(\frac{d\varphi'}{d\alpha} \right)_0 &= 0 \\ \cos\theta\sin\varphi \left(\frac{d\theta'}{d\alpha} \right)_0 - \sin\theta\cos\varphi \left(\frac{d\varphi'}{d\alpha} \right)_0 &= \cos\theta \\ -\sin\theta \left(\frac{d\theta'}{d\alpha} \right)_0 &= -\sin\theta\sin\varphi \end{aligned}\right\} \quad (12.9)$$

From these relationships we readily obtain

$$\left.\begin{aligned} a_1(\theta,\ \varphi) &= \left(\frac{d\theta'}{d\alpha}\right)_0 = \sin\varphi \\ b_1(\theta,\ \varphi) &= \left(\frac{d\varphi'}{d\alpha}\right)_0 = \cot\theta\cos\varphi \end{aligned}\right\} \qquad (12.10)$$

and, consequently,

$$\hat{A}_1 f(\theta,\ \varphi) = \sin\varphi\frac{\partial f}{\partial\theta} + \cot\theta\cos\varphi\frac{\partial f}{\partial\varphi} \qquad (12.11)$$

Similarly,

$$\hat{A}_2 f = -\cos\varphi\frac{\partial f}{\partial\theta} + \cot\theta\sin\varphi\frac{\partial f}{\partial\varphi} \qquad (12.12)$$

Let us now find the basis of an irreducible representation of weight j in the space of the functions $f(\theta,\ \varphi)$. We know that any element of the basis of an irreducible representation of the rotation group must satisfy the equation

$$(\hat{A}_1^2 + \hat{A}_2^2 + \hat{A}_3^2) f(\theta,\ \varphi) = -j(j+1) f(\theta,\ \varphi) \qquad (12.13)$$

If we wish to find the canonical basis we must, in addition, demand that

$$i\hat{A}_3 f(\theta,\ \varphi) = mf(\theta,\ \varphi), \quad m = -j,\ -j+1,\ \ldots,\ j \qquad (12.14)$$

Substituting (12.3), (12.11) and (12.12) in (12.13) and (12.14), we obtain

$$\frac{1}{\sin\theta}\frac{\partial}{\partial\theta}\left(\sin\theta\frac{\partial f}{\partial\theta}\right) + \frac{1}{\sin^2\theta}\frac{\partial^2 f}{\partial\varphi^2} + j(j+1)f = 0 \qquad (12.15)$$

$$-i\frac{\partial f}{\partial\varphi} = mf \qquad (12.16)$$

These two equations are the equations for spherical functions, and have a solution only for integral values $j = l$. The number of linearly independent solutions of these equations is $2l+1$ and, consequently, this number is equal to the order of the irreducible representation $D^{(l)}$. Therefore, the unit vectors of the canonical basis of an irreducible representation with an integral weight l in the space of the continuous functions $f(\theta,\ \varphi)$ are of the form

$$Y_l^m(\theta,\ \varphi) = \frac{1}{\sqrt{2\pi}} e^{im\varphi} P_l^m(\cos\theta) \qquad (12.17)$$

where P_l^m is the normalized associated Legendre polynomial

$$P_l^m(x) = (-1)^m \sqrt{\frac{(l-|m|)!}{(l+|m|)!}} \sqrt{\frac{2l+1}{2}} \frac{1}{2^l l!} \times (1-x^2)^{\frac{|m|}{2}} \frac{d^{l+|m|}(x^2-1)^l}{dx^{l+|m|}}$$

$$(P_l^{-m}(x) = P_l^m(x)) \qquad (12.18)$$

Let us now write down the transformation for the spherical functions under rotation. If we know the infinitesimal matrices A_i of the irreducible representation, we can find the matrices of this representation for arbitrary rotations. However, simpler expressions are obtained if we take the Euler angles $\varphi_1, \theta_1, \varphi_2$ as the rotation parameters. We shall not go into details of the derivation and will merely quote the final result. Suppose that the rotation $g[\varphi_1, \theta_1, \varphi_2]$ transforms the vector $\boldsymbol{n}'(\theta', \varphi')$ into the vector $\boldsymbol{n}(\theta, \varphi)$, i.e. $g^{-1}\boldsymbol{n}'(\theta', \varphi') = \boldsymbol{n}(\theta, \varphi)$. Then,

$$\hat{T}_g Y_l^m(\theta, \varphi) = Y_l^m(\theta', \varphi') = \sum_{m'=-l}^{l} D_{mm'}^{(l)}[\varphi_1, \theta_1, \varphi_2] Y_l^{m'}(\theta, \varphi) \qquad (12.19)$$

where

$$D_{mm'}^{(l)}[\varphi_1, \theta_1, \varphi_2] = \sqrt{\frac{(l+m)!(l-m)!}{(l+m')!(l-m')!}} e^{i(m\varphi_1 + m'\varphi_2)} \times \left[\cos\frac{1}{2}\theta\right]^{m+m'} \left[\sin\frac{1}{2}\theta\right]^{m-m'} P_{l-m}^{(m-m', m+m')}(\cos\theta) \qquad (12.20)$$

and $P_n^{(a,b)}(\cos\theta)$ are the Jacobi polynomials

$$P_n^{(a,b)}(x) = \frac{(-1)^n}{2^n n!}(1-x)^{-a}(1+x)^{-b} \frac{d^n}{dx^n} \times [(1-x)^{a+n}(1+x)^{b+n}]$$

12.2 Composition of the irreducible representations of the group $O^+(3)$

Consider two irreducible representations $D^{(j)}$ and $D^{(j')}$ of the group $O^+(3)$, which are realized in the spaces R_{2j+1} and $R_{2j'+1}$.

We shall denote the canonical bases of these representations by

$$u_k^{(j)}\ (k=-j,\ -j+1,\ \ldots,\ j)$$

and

$$v_m^{(j')}\ (m=-j',\ -j'+1,\ \ldots,\ j')$$

respectively. Let us now form the composition $D^{(j)}\times D^{(j')}$ of the two irreducible representations. The resultant representation will be realized in the space $R_{(2j+1)(2j'+1)}$ which is the direct product of the two spaces R_{2j+1} and $R_{2j'+1}$. For the basis vectors in the space $R_{(2j+1)(2j'+1)}$ we shall take all the possible products $u_k^{(j)}v_m^{(j')}$, which we shall denote by $w_{km}^{(jj')}$. The transformation of the basis vectors $w_{km}^{(jj')}$ is defined by the formula

$$\hat{T}_g w_{km}^{(jj')}=\sum_{s=-j}^{j} D_{sk}^{(j)}(g)\,u_s^{(j)}\sum_{r=-j'}^{j'} D_{rm}^{(j')}(g)\,v_r^{(j')}$$

$$=\sum_{s,r}\left\{D^{(j)}(g)\times D^{(j')}(g)\right\}_{sr,\,km} w_{sr}^{(jj')} \tag{12.21}$$

Let us find the infinitesimal matrices of the direct product. Expanding each of the matrices $D^{(j)}$ and $D^{(j')}$ in powers of the parameters α_i, we obtain

$$D^{(j)}\times D^{(j')}=(E_{2j+1}+\alpha_1 A_1^{(j)}+\ldots)$$
$$\times(E_{2j'+1}+\alpha_1 A_1^{(j')}+\ldots)=E_{2j+1}\times E_{2j'+1}$$
$$+\alpha_1(A_1^{(j)}\times E_{2j'+1}+E_{2j+1}\times A_1^{(j')})+\ldots \tag{12.22}$$

from which it follows that the infinitesimal matrices of our representation are of the form

$$A_i^{(jj')}=A_i^{(j)}\times E_{2j'+1}+E_{2j+1}\times A_i^{(j')} \tag{12.23}$$

We shall now show that the basis vectors $w_{km}^{(jj')}$ of the space $R_{(2j+1)(2j'+1)}$ will be the eigenvectors of the matrix $H_3^{(jj')}=iA_3^{(jj')}$ with the eigenvalues $k+m$. In fact,

$$\left.\begin{aligned} H_3^{(jj')}w_{km}&=(H_3^{(j)}\times E_{2j'+1}+E_{2j+1}\times H_3^{(j')})\,w_{km}\\ &H_3^{(j)}u_k\times E_{2j'+1}v_m+E_{2j+1}u_k\times H_3^{(j')}v_m\\ =ku_kv_m+u_kmv_m&=(k+m)\,u_kv_m=(k+m)\,w_{km}\end{aligned}\right\} \tag{12.24}$$

Table of eigenvalues and eigenvectors of the matrix $H_3^{(jj')}$ $(j' \geq j)$

Eigenvalues of $H_3^{(jj')}$	Eigenvectors of $H_3^{(jj')}$ [1]	Multiplicity
$j+j'$	$w_{jj'}$	1
$j+j'-1$	$w_{j-1\,j'}$, $w_{j\,j'-1}$	2
. . . . .	. . . .	. . .
$j'-j+1$	$w_{-j+1\,j'}$, $w_{-j+2\,j'-1}$, ..., $w_{j\,j'-2j+1}$	$2j$
$j'-j$	$w_{-j\,j'}$, $w_{-j+1\,j'-1}$, ..., $w_{j\,j'-2j}$	$2j+1$
$j'-j-1$	$w_{-j\,j'-1}$, $w_{-j+1\,j'-2}$, ..., $w_{j\,j'-2j-1}$	$2j+1$
. . . . .	. . . .	. . .
$-(j'-j)$	$w_{-j\,-j'+2j}$, $w_{-j+1\,-j'+2j-1}$, ..., $w_{j\,-j'}$	$2j+1$
$-(j'-j)-1$	$w_{-j\,-j'+2j-1}$, ..., $w_{j-1\,-j'}$	$2j$
. . . .	. . . .	. . .
$-j-j'$	$w_{-j\,-j'}$	1

[1] The superscripts of the vectors $w_{km}^{(jj')}$ are omitted for the sake of brevity.

Since the number k can assume the $2j+1$ values $-j, \ldots, j$, and the number m can assume the $2j'+1$ values $-j', \ldots, j'$, it follows that their sum $M=k+m$ can assume the $2(j+j')+1$ different values $-j'-j \leqslant M \leqslant j+j'$. Since the number of basis vectors $w_{km}^{(jj')}$ is equal to $(2j+1)(2j'+1)$, we conclude that some of the eigenvalues M can be degenerate (if j or j' is not zero). Hence, it follows that with the exception of the case when $j=0$ or $j'=0$, the representation $D^{(j)} \times D^{(j')}$ is reducible and, therefore, can be decomposed into irreducible parts. This is called the Clebsch-Gordan decomposition (see Chapter 4) and can be obtained by dividing the space $R_{(2j+1)(2j'+1)}$ into invariant orthogonal sub-spaces in which one of the irreducible representations of the rotation group is realized. This can be done with the aid of the table opposite, from which one can determine the degree of degeneracy of the eigenvalues of the operator $H_3^{(jj')}$. To be specific we shall suppose that $j' \geqslant j$. Consider the greatest eigenvalue of the matrix $H_3^{(jj')}$. It is not degenerate and is equal to $j+j'$. If we apply the successive powers of the matrix $H_-^{(jj')}$ to the basis vector $w_{jj'}^{(jj')}$, we obtain the following chain of eigenvectors of the matrix $H_3^{(jj')}$:

$$w_{jj'}^{(jj')},\ H_-^{(jj')} w_{jj'}^{(jj')},\ \ldots,\ \left(H_-^{(jj')}\right)^{2(j'+j)} w_{jj'}^{(jj')} \qquad (12.25)$$

According to Section 11.2, these vectors should form the basis of an irreducible representation of weight $J=j+j'$. The space in which the representation is realized will be denoted by $R_{2(j+j')+1}$.

Consider now the orthogonal complement of the space $R_{2(j+j')+1}$ in $R_{(2j+1)(2j'+1)}$. It follows from the table that in this complement there is only one vector with the eigenvalue $j+j'-1$. If we apply to this vector the successive powers of the matrix $H_-^{(jj')}$, we obtain a chain of $2(j+j'-1)+1$ vectors. These vectors form the basis of the irreducible representation $D^{(j+j'-1)}$. In the orthogonal complement of the space $R_{2(j+j'-1)+1}$ we shall have one vector with the maximum eigenvalue $j+j'-2$. If we repeat our analysis, we can finally represent the space $R_{(2j+1)(2j'+1)}$ in the form of the direct sum of the orthogonal sub-space

$$R_{(2j+1)(2j'+1)} = R_{2(j+j')+1} \oplus R_{2(j+j'-1)+1} \oplus \dots \oplus R_{2(j'-j)} \quad (12.26)$$

Consequently, the Clebsch-Gordan decomposition of the group $O^+(3)$ is of the form

$$D^{(j)} \times D^{(j')} = D^{(j+j')} \oplus D^{(j+j'-1)} \oplus \dots \oplus D^{(|j-j'|)} \quad (12.27)$$

We thus have the following result: the direct product of irreducible representations with weights j and j' decomposes into irreducible representations with weights $j+j'$, $j+j'-1, \dots, |j-j'|$, and each of these irreducible representations is encountered in the decomposition only once.

From this rule it follows immediately that the identity representation enters the representation $D^{(j)} \times D^{(j')}$ only when $j=j'$. From (12.27) it follows that the product of two two-valued representations (j and j' are half-integers) is a single-valued representation.

The basis vectors $\boldsymbol{w}_M^{(J)}$ of the canonical basis, in which the representation $D^{(j)} \times D^{(j')}$ decomposes into irreducible representations, are linear combinations of the vectors $\boldsymbol{w}_{km}^{(jj')} = \boldsymbol{u}_k^{(j)} \boldsymbol{v}_m^{(j')}$:

$$\boldsymbol{w}_M^{(J)} = \sum_{m,\, m'} c_{mm'M}^{(jj'J)} \boldsymbol{u}_m^{(j)} \boldsymbol{v}_m^{(j')} \quad (12.28)$$

The coefficients $c_{mm'M}^{(jj'J)}$ are called the Clebsch-Gordan or Wigner coefficients for the group $O^+(3)$. The matrix $C^{(jj')}$ with the elements $c_{mm'M}^{(jj'J)}$ (the indices J, M label the rows and m, m' the columns) is a unitary matrix of order $(2j+1)(2j'+1)$. The matrix $C^{(jj')}$ reduces the representation $D^{(j)} \times D^{(j')}$ to the quasi-diagonal form

$$(C^{(jj')})^{-1} (D^{(j)} \times D^{(j')}) C^{(jj')}$$

$$
= \begin{pmatrix} D^{(j+j')} & 0 & \dots & 0 \\ 0 & D^{(j+j'-1)} & \dots & 0 \\ & \cdot & & \\ & & \cdot & \\ & & & \cdot \\ 0 & 0 & \dots & D^{(|j-j'|)} \end{pmatrix} \tag{12.29}
$$

In some problems it is necessary to consider the product of three irreducible representations $D^{(j_1)} \times D^{(j_2)} \times D^{(j_3)}$ which is realized in the space $R = R_{j_1} \times R_{j_2} \times R_{j_3}$. To construct the canonical basis in this space, we can start with the canonical basis vectors $w^{(j_{12})}_{m_{12}}$ in the space $R_{j_1} \times R_{j_2}$, and then combine them with the vectors $w^{(j_3)}_{m_3}$ of the space R_{j_3}. The basis vectors obtained in this way will be denoted by $w^{((j_{12})\,j)}_{m}$. However, it is possible to proceed in a different way. Let us first construct the basis vectors $w^{(j_{23})}_{m_{23}}$ and then combine them with the basis vectors $w^{(j_1)}_{m_1}$. As a result we obtain $w^{((j_{23})\,j)}_{m}$. It is clear that $w^{((j_{12})\,j)}_{m}$ are linear combinations of the vectors $w^{((j_{23})\,j)}_{m}$:

$$
w^{((j_{12})\,j)}_{m} = \sum_{j_{23}} \langle j_1 j_2 (j_{12})\, j_3 j \,|\, j_1 j_2 j_3 (j_{23})\, j \rangle\, w^{((j_{23})\,j)}_{m} \tag{12.30}
$$

The coefficients $\langle j_1 j_2 (j_{12})\, j_3 j \,|\, j_1 j_2 j_3 (j_{23})\, j \rangle$ are called the Racah coefficients. It can be shown that

$$
\langle j_1 j_2 (j_{12})\, j_3 j \,|\, j_1 j_2 j_3 (j_{23})\, j \rangle
$$

$$
= \sum_{m_1+m_2+m_3=m} C^{(j_1 j_2 j_{12})}_{m_1 m_2 m_{12}} C^{(j_{12} j_3 j)}_{m_{12} m_3 m} C^{(j_2 j_3 j_{23})}_{m_2 m_3 m_{23}} C^{(j_1 j_{23} j)}_{m_1 m_{23} m} \tag{12.31}
$$

12.3 Tensor and spinor representations of the rotation group

Let us now introduce the concept of a tensor of rank n in

three-dimensional space. We shall say that we have a given tensor of rank n if in each orthogonal set of coordinates we have defined a set of 3^n numbers $T_{i_1 i_2 \ldots i_n}$ which obey the following transformation rule:

$$T'_{i'_1 i'_2 \ldots i'_n} = \sum_{i_1=1}^{3} \sum_{i_2=1}^{3} \cdots \sum_{i_n=1}^{3} g_{i'_1 i_1} g_{i'_2 i_2} \cdots g_{i'_n i_n} T_{i_1 i_2 \ldots i_n} \qquad (12.32)$$

where the matrix $\| g_{ik} \|$ relates the unit vectors of the old and new sets of coordinates:

$$e'_i = \sum_{k=1}^{3} g_{ki} e_k \qquad (12.33)$$

We see that the transformation matrix for the components of a tensor is identical with the matrix of the representation which is the direct product of n 'vector' representations. We shall call this the tensor representation of rank n. The tensor representations are, of course, reducible. They can be decomposed into irreducible representations by the Clebsch-Gordan rule. The tensor representation of any rank is single-valued.

Similarly, we can introduce the concept of a spinor of rank n, and a spinor representation of the rotation group. While in the definition of the tensor representation the basic relation was (12.33), the definition of the spinor representation is based on the two-valued irreducible representation $D^{(1/2)}$. The matrix of this representation expressed in terms of the Euler angles is of the form (see Equation (11.38))

$$\begin{pmatrix} a_{11} & a_{12} \\ a_{21} & a_{22} \end{pmatrix} = \begin{pmatrix} \alpha & \beta \\ -\bar{\beta} & \bar{\alpha} \end{pmatrix} \qquad (12.34)$$

where

$$\alpha = \pm \cos\frac{\theta}{2} e^{i\frac{\varphi_1+\varphi_2}{2}}, \quad \beta = \mp i \sin\frac{\theta}{2} e^{i\frac{\varphi_2-\varphi_1}{2}}$$

We shall say that we have spinor of rank n if in each orthogonal set of coordinates we have defined a set of 2^n

complex numbers $\chi_{\lambda_1\lambda_2\ldots\lambda_n}$ ($\lambda_i = 1, 2$) which transform in accordance with the rule

$$\chi'_{\lambda'_1\lambda'_2\ldots\lambda'_n} = \sum_{\lambda_1,\ldots,\lambda_n} a_{\lambda'_1\lambda_1} a_{\lambda'_2\lambda_2} \ldots a_{\lambda'_n\lambda_n} \chi_{\lambda_1\lambda_2\ldots\lambda_n} \qquad (12.35)$$

This formula defines the transformation rule for a contravariant spinor. In addition to contravariant spinors we can also define covariant spinors for which the transformation rule is obtained from (12.35) by replacing $a_{\lambda'_i\lambda_i}$ by their complex conjugates (see Section 12.4).

The above transformation rule defines the representation of the rotation group which is the direct product of irreducible representations $D^{(1/2)}$. We shall call this the spinor representation of rank n. The decomposition of a spinor representation into irreducible representations can also be performed by the Clebsch-Gordan rule. It is clear that spinors of even rank transform by single-valued representations, while spinors of odd rank transform by two-valued representations.

Since the transformation matrix (12.35) can be written as the direct product of n matrices $\|a_{ik}\|$ of the representation $D^{(1/2)}$:

$$\|a_{\lambda'_1\lambda_1}\| \times \|a_{\lambda'_2\lambda_2}\| \times \ldots \times \|a_{\lambda'_n\lambda_n}\| \qquad (12.36)$$

it follows that the infinitesimal matrices of the spinor representation can be written in the form (cf. Equation (12.23))

$$A_i = \sum E_2 \times E_2 \times \ldots \times E_2 \times A_i^{(1/2)} \times E_2 \times \ldots \times E_2 \qquad (12.37)$$

where $A_i^{(1/2)}$ are the infinitesimal matrices of the irreducible representation $D^{(1/2)}$, and the sum is evaluated over all the positions of this factor in the product.

Consider the spinor $\chi_{\lambda_1\lambda_2\ldots\lambda_n}$ of which the only non-zero component in the given set of coordinates is $\chi_{\lambda_1\lambda_2\ldots\lambda_n}$. Let us establish the effect of the infinitesimal matrix $H_3 = iA_3$ on the spinor. We recall that the matrix

$iA_3^{(1/2)}$ is given by

$$iA_3^{(1/2)} = \begin{pmatrix} \frac{1}{2} & 0 \\ 0 & -\frac{1}{2} \end{pmatrix} \qquad (12.38)$$

and, therefore,

$$H_3 \chi_{\lambda_1 \lambda_2 \ldots \lambda_n} = \left(\frac{p}{2} - \frac{q}{2}\right) \chi_{\lambda_1 \lambda_2 \ldots \lambda_n} \qquad (12.39)$$

where p is the number of indices of a given spinor component equal to unity, and q is the number of indices equal to 2. If instead of the indices λ_i we use the indices $\sigma_i = \frac{1}{2}, -\frac{1}{2}$, we have

$$H_3 \chi_{\sigma_1 \sigma_2 \ldots \sigma_n} = (\sigma_1 + \sigma_2 + \ldots + \sigma_n) \chi_{\sigma_1 \sigma_2 \ldots \sigma_n} \qquad (12.40)$$

The spinor $\chi_{\sigma_1 \sigma_2 \ldots \sigma_n}$ is thus the eigenspinor of the matrix H_3, which corresponds to the eigenvalue equal to the sum of the spinor indices σ_i.

12.4 Complex conjugate representations

We shall say that we have defined a j-vector if we can associate with each orthogonal set of coordinates the set of $2j+1$ numbers ξ_m, $m = -j, -j+1, \ldots, j$ which transform in accordance with the rule

$$\xi'_m = \sum_{m'=-j}^{j} D^{(j)}_{mm'}(\alpha_1, \alpha_2, \alpha_3)\, \xi_{m'} \qquad (12.41)$$

where $\| D^{(j)}_{mm'} \|$ are the matrices of an irreducible representation. Let us associate with each set of coordinates the set of complex conjugates $\bar{\xi}'_m$, where ξ'_m are defined by (12.41). It is clear that $\bar{\xi}'_m$ and $\bar{\xi}_m$ are related through complex conjugate matrices:

$$\bar{\xi}'_m = \sum_{m'=-j}^{j} \bar{D}^{(j)}_{mm'} \bar{\xi}_{m'} \qquad (12.42)$$

The matrices $\bar{D}^{(j)}$, like the matrices $D^{(j)}$, form a representation of the rotation group. Since the representation $D^{(j)}$ is irreducible, it follows that the representation $\bar{D}^{(j)}$ is also irreducible. Since the representations are of the same order, they should be equivalent:

$$\bar{D}^{(j)} = VD^{(j)}V^{-1} \tag{12.43}$$

The matrix V is defined by this equation to within a scalar factor. To show this, let us suppose that there are two matrices V and W such that

$$\bar{D}^{(j)} = VD^{(j)}V^{-1} = WD^{(j)}W^{-1}$$

We can then write

$$D^{j}V^{-1}W = V^{-1}WD^{(j)}$$

and therefore

$$V^{-1}W = \lambda E \tag{12.44}$$

i.e. $V^{-1}W$ is a multiple of the unit matrix. Consequently,

$$W = \lambda V \tag{12.45}$$

Equation (12.43) shows that the components $\bar{\xi}_m$ of the complex conjugate vector transform in accordance with the same rule as the linear combinations $\sum v_{ik}\xi_k$ of the initial j-vector. To find the matrix V let us write (12.43), retaining only linear terms in α_i:

$$E + \alpha_1\bar{A}_1 + \alpha_2\bar{A}_2 + \alpha_3\bar{A}_3$$

$$= E + \alpha_1 VA_1V^{-1} + \alpha_2 VA_2V^{-1} + \alpha_3 VA_3V^{-1} \tag{12.46}$$

Hence it follows that

$$\bar{A}_i = VA_iV^{-1} \qquad (i = 1,\ 2,\ 3) \tag{12.47}$$

and if we use the explicit form of $A_i^{(j)}$ given by (11.28), we have

$$\left.\begin{aligned} -A_1^{(j)} &= VA_1^{(j)}V^{-1} \\ A_2^{(j)} &= VA_2^{(j)}V^{-1} \\ -A_3^{(j)} &= VA_3^{(j)}V^{-1} \end{aligned}\right\} \tag{12.48}$$

It is clear that we can define V as the matrix of the representation of rotation through 180° about the y-axis. Consequently,

$$V = e^{A_2 \pi} \tag{12.49}$$

Let us find an expression for V for the representation $D^{(1/2)}$. Using the property $A_2^{(1/2)}$:

$$(2iA_2^{(1/2)})^2 = \begin{pmatrix} 0 & -i \\ i & 0 \end{pmatrix}^2 = \begin{pmatrix} 1 & 0 \\ 0 & 1 \end{pmatrix} \tag{12.50}$$

and expanding the exponential in (12.49) into a series, we find that

$$V^{(1/2)} = e^{-i(2iA_2^{(1/2)})\frac{\pi}{2}} = \cos\frac{\pi}{2} - \left(i\sin\frac{\pi}{2}\right)(2iA_2^{(1/2)}) = 2A_2^{(1/2)}$$

and therefore

$$V^{(1/2)} = \begin{pmatrix} 0 & -1 \\ 1 & 0 \end{pmatrix} \tag{12.51}$$

Suppose that we have a given spinor $\begin{pmatrix} a \\ b \end{pmatrix}$. In view of the foregoing discussion, we may conclude that the quantity

$$\begin{pmatrix} \tilde{a} \\ \tilde{b} \end{pmatrix} = (2A_2^{1/2})^{-1}\begin{pmatrix} \bar{a} \\ \bar{b} \end{pmatrix} = \begin{pmatrix} 0 & 1 \\ -1 & 0 \end{pmatrix}\begin{pmatrix} \bar{a} \\ \bar{b} \end{pmatrix} = \begin{pmatrix} \bar{b} \\ -\bar{a} \end{pmatrix} \tag{12.52}$$

where $\begin{pmatrix} \bar{a} \\ \bar{b} \end{pmatrix}$ is the complex conjugate spinor, transforms as a spinor. Conversely, the quantity

$$(2A_2^{(1/2)})\begin{pmatrix} a \\ b \end{pmatrix} = \begin{pmatrix} -b \\ a \end{pmatrix} \tag{12.53}$$

transforms as a complex conjugate spinor. A complex conjugate spinor is called a covariant spinor.

Exercises

12.1. Prove (12.31) using the unitarity of the matrix $C^{(jj')}$.

12.2. Find the decomposition into irreducible representations of a tensor representation of rank 3 and 4.

12.3. Show that the decomposition of a tensor representation into irreducible representations can be found with the aid of the following recurrence relation:

$$\varkappa_n^{(l)} = \varkappa_{n-1}^{(l-1)} + \varkappa_{n-1}^{(l)} + \varkappa_{n-1}^{(l+1)}$$

where $\varkappa_n^{(l)}$ is a number showing how many times an irreducible representation of weight l is contained in the tensor representation of rank n.

12.4. Show that the decomposition of a spinor representation into irreducible parts can be obtained with the aid of the following recurrence relation:

$$\eta_n^{(j)} = \eta_{n-1}^{\left(j-\frac{1}{2}\right)} + \eta_{n-1}^{\left(j+\frac{1}{2}\right)}$$

where $\eta_n^{(j)}$ is a number showing how many times an irreducible representation of weight j is present in the spinor representation of rank n. Using this formula, find the decompositions of spinor representations for $n = 1, 2, \ldots, 6$.

Chapter 13

Some Applications of the Theory of Representation of the Rotation Group in Quantum Mechanics

13.1 Particle in a central force field. Angular momentum

Consider the Schroedinger equation for a particle in a central force field:

$$\left\{-\frac{\hbar^2}{2m}\Delta + V(r)\right\}\psi(\boldsymbol{r}) = E\psi(\boldsymbol{r}) \tag{13.1}$$

Since the potential $V(r)$ is a function of only the modulus of the position vector, the Hamiltonian $\hat{H}$ for this system is invariant under the group $O^+(3)$. We know (see Chapter 5) that this invariance condition can be written in the form

$$\hat{T}_g\hat{H} = \hat{H}\hat{T}_g \tag{13.2}$$

where the operators $\hat{T}_g$, which form the representation of the rotation group, are defined by

$$\hat{T}_g\psi(\boldsymbol{r}) = \psi(g^{-1}\boldsymbol{r}) \tag{13.3}$$

Let us find the infinitesimal operators $\hat{A}_1, \hat{A}_2, \hat{A}_3$ of this representation which correspond to rotations about the axes x_1, x_2, x_3. The infinitesimal operators $\hat{A}_i$ define the increment of the function $\psi(\boldsymbol{r})$ due to the rotation of the position vector which is linear in $\alpha_1, \alpha_2, \alpha_3$. We can therefore write

$$\delta_i\psi = \alpha_i\hat{A}_i\psi = \sum_{\mu=1}^{3} \delta_i x_\mu \frac{\partial}{\partial x_\mu}\psi \tag{13.4}$$

where, according to (10.25),

$$\left.\begin{array}{lll} \delta_1 x_1 = 0, & \delta_1 x_2 = \alpha_1 x_3, & \delta_1 x_3 = -\alpha_1 x_2 \\ \delta_2 x_1 = -\alpha_2 x_3, & \delta_2 x_2 = 0, & \delta_2 x_3 = \alpha_2 x_1 \\ \delta_3 x_1 = \alpha_3 x_2, & \delta_3 x_2 = -\alpha_3 x_1, & \delta_3 x_3 = 0 \end{array}\right\} \tag{13.5}$$

Hence, we find that

$$\left.\begin{array}{l} \hat{A}_1 = x_3 \dfrac{\partial}{\partial x_2} - x_2 \dfrac{\partial}{\partial x_3} \\ \hat{A}_2 = -x_3 \dfrac{\partial}{\partial x_1} + x_1 \dfrac{\partial}{\partial x_3} \\ \hat{A}_3 = x_2 \dfrac{\partial}{\partial x_1} - x_1 \dfrac{\partial}{\partial x_2} \end{array}\right\} \tag{13.6}$$

If we now retain only the linear terms, the operator $\hat{T}_g$ can be written in the form

$$\hat{T}_g = \hat{E} + \alpha_1\hat{A}_1 + \alpha_2\hat{A}_2 + \alpha_3\hat{A}_3 \tag{13.7}$$

and hence the commutation rule (13.2) yields

$$\hat{A}_i\hat{H} = \hat{H}\hat{A}_i \tag{13.8}$$

Conversely, since the operator $\hat{T}_g$ can, according to (12.2), be written in the form of the series

$$\begin{aligned} \hat{T}_g &= \exp(\alpha_1\hat{A}_1 + \alpha_2\hat{A}_2 + \alpha_3\hat{A}_3) \\ &= \hat{E} + (\alpha_1\hat{A}_1 + \alpha_2\hat{A}_2 + \alpha_3\hat{A}_3) \\ &+ \frac{1}{2}(\alpha_1\hat{A}_1 + \alpha_2\hat{A}_2 + \alpha_3\hat{A}_3)^2 + \dots \end{aligned} \tag{13.9}$$

the result given by (13.2) follows from (13.8). Consequently, the equations in (13.8) are the necessary and sufficient conditions for the invariance of the Hamiltonian under the rotation group. The operators $\hat{A}_j$ are anti-Hermitian. Let us introduce the Hermitian operators $\hat{H}_j = i\hat{A}_j$, which can be regarded as the quantum-mechanical operators corresponding to certain physical observables. In fact, the operators $\hat{H}_j$ are identical, apart from the dimensional factor $\hbar$, with the operators $\hat{L}_j$ which represent the components of the angular momentum, for example,

$$\hat{L}_1 = \hbar\hat{H}_1 = i\hbar\hat{A}_1 = i\hbar\left(x_3\frac{\partial}{\partial x_2} - x_2\frac{\partial}{\partial x_3}\right) = x_2\hat{p}_3 - x_3\hat{p}_2 \qquad (13.10)$$

where $\hat{p}_j$ is a component of the momentum operator $p_j = -i\hbar\frac{\partial}{\partial x_j}$. In view of (13.8) the operators $\hat{L}_j$ commute with the Hamiltonian $\hat{H}$ for our system and, consequently, they are constants of motion.

We have thus shown that the operators representing the components of the angular momentum are identical, apart from a factor, with the infinitesimal operators of the representation of the rotation group, and if the Schroedinger equation is invariant under this group, they are constants of motion.

Consider now the symmetry properties of the solutions of (13.1). The eigenfunctions of this equation which belong to a given energy eigenvalue should form the basis of an irreducible representation of the rotation group. We know from the last chapter that the only representations which can be realized in the space of these functions are $D^{(l)}$, where l is an integer. Moreover, functions transforming in accordance with the irreducible representation $D^{(l)}$ should satisfy the equation

$$(\hat{H}_1^2 + \hat{H}_2^2 + \hat{H}_3^2)\psi(\boldsymbol{r}) = l(l+1)\psi(\boldsymbol{r})$$

or

$$(\hat{L}_1^2 + \hat{L}_2^2 + \hat{L}_3^2)\psi(\boldsymbol{r}) = \hbar^2 l(l+1)\psi(\boldsymbol{r}) \qquad (13.11)$$

If the functions corresponding to a degenerate eigenvalue form a canonical basis of the representation $D^{(l)}$, they will also satisfy the equation

$$\hat{H}_3\psi(\boldsymbol{r}) = m\psi(\boldsymbol{r}) \qquad (13.12)$$

or

$$\hat{L}_3\psi(\boldsymbol{r}) = \hbar m\psi(\boldsymbol{r}) \qquad (m = -l, -l+1, \ldots, l)$$

The explicit form of functions forming the basis of an irreducible representation of the rotation group was obtained in Chapter 12. We may therefore conclude that the solutions of the Schroedinger equation (13.1) are of the form

$$\psi_{lm}(\boldsymbol{r}) = R(r)\,Y_l^{(m)}(\theta, \varphi) \tag{13.13}$$

We have thus reached the following result. The eigenfunctions of the Schroedinger equation (13.1) corresponding to a given eigenvalue form the basis of an irreducible representation $D^{(l)}$ of the rotation group, where l is an integer. The degree of degeneracy of a level E_l is $2l+1$. The corresponding eigenfunctions are also the eigenfunctions of the operator representing the square of the angular momentum. They can always be chosen so that they are also the eigenfunctions of the operator $\hat{L}_3$.

We know further that the complete symmetry group for the problem in hand is the group $O(3) = O^+(3) \times \boldsymbol{i}$. Depending on whether the wave function is symmetric with respect to the inversion $\boldsymbol{i}$ or changes sign, the corresponding state must be ascribed the parity quantum number $w = \pm 1$. In view of (13.13), it is clear that for a single particle

$$w = (-1)^l$$

To prove this, it is sufficient to recall the relation between the spherical harmonics Y_{lm} and homogeneous polynomials of order l of the variables x, y, z. It follows that, for a single particle, the classification based on the eigenvalues of the orbital angular momentum already contains the parity classification. We shall see later that this is not so for multi-electron systems (Chapter 19).

13.2 Addition of angular momenta

We showed in the last section that the operators representing the components of the angular momentum are identical, apart from a factor, with the infinitesimal operators of the rotation group. We can use this relation to establish the addition rule for angular momenta.

Suppose we have two non-interacting particles, each moving in a central force field, i.e. each described by the Schroedinger equation (13.1). The wave functions of the particles will be denoted by $\varphi_{l_1m_1}(\boldsymbol{r}_1)$ and $\psi_{l_2m_2}(\boldsymbol{r}_2)$, respectively. In order to avoid complications by having to take into account the identity principle (Chapter 17), we shall suppose that the particles are quite different (one is an electron and the other a proton). Possible wave functions for the system can then be written in the form of the product

$$\varphi_{l_1m_1}(\boldsymbol{r}_1)\,\psi_{l_2m_2}(\boldsymbol{r}_2) \tag{13.14}$$

Our problem is to determine the possible eigenvalues $\hbar^2L(L+1)$ of the operator representing the square of the total angular momentum $\hat{L}^2=(\hat{L}^{(1)}+\hat{L}^{(2)})^2$ for the states (13.14). The effect of the operator $\hat{L}^{(1)}+\hat{L}^{(2)}$ is defined by

$$(\hat{L}^{(1)}+\hat{L}^{(2)})\,\varphi_{l_1m_1}(\boldsymbol{r}_1)\,\psi_{l_2m_2}(\boldsymbol{r}_2)=\psi_{l_2m_2}\hat{L}^{(1)}\varphi_{l_1m_1}+\varphi_{l_1m_1}\hat{L}^{(2)}\psi_{l_2m_2} \tag{13.15}$$

Comparison of this with (12.23) will show that, in the present case, the components of the total angular-momentum operator are the same (apart from a factor) as the infinitesimal operators of the direct product of the representations $D^{(l_1)}$ and $D^{(l_2)}$. Our problem is thus simply reduced to the decomposition of the direct product of two irreducible representations of the rotation group into irreducible representations. Using the Clebsch-Gordan rule, we find that the quantum number L can assume the values l_1+l_2, $l_1+l_2-1, \ldots, |l_1-l_2|$ and, according to (12.28), the eigenfunctions of the operators $\hat{L}^2$ and $\hat{L}$ are of the form

$$\Psi_{LM}(\boldsymbol{r}_1, \boldsymbol{r}_2)=\sum_{m_1,\, m_2} c^{(l_1l_2L)}_{m_1m_2M}\varphi_{l_1m_1}(\boldsymbol{r}_1)\,\psi_{l_2m_2}(\boldsymbol{r}_2) \tag{13.16}$$

where $c^{(l_1l_2L)}_{m_1m_2M}$ are the Clebsch-Gordan coefficients.

13.3 Spin

In Section 13.1 we used the solutions of (13.1) which transform by single-valued representations of the rotation group to describe the states of a particle in a central field. The question as to whether such solutions, and indeed the equation itself, are capable of describing real physical situations can only be settled by comparison with experiment. Experiment does, in fact, show that (13.1) cannot explain some of the observed properties of the electron. In particular, it has been found that violation of spherical symmetry by the inclusion of an external magnetic field leads to the splitting of the ground-state energy E_0 $(l=0)$, which should not be degenerate according to the theory developed in Section 13.1. This contradiction is removed, however, if we suppose that the electron wave function transforms under rotations in accordance with a two-valued representation of the group $O^+(3)$.

In non-relativistic quantum mechanics the wave function of an electron is assumed to be the two-component

$$\begin{pmatrix} \chi_1(\boldsymbol{r}) \\ \chi_2(\boldsymbol{r}) \end{pmatrix}$$

quantity. Under rotations of the system of coordinates this quantity transforms by the rule

$$\chi_i'(\boldsymbol{r}') = \sum_j a_{ij}(g)\chi_j(\boldsymbol{r}) \quad (\boldsymbol{r}' = g\boldsymbol{r}) \tag{13.17}$$

where $\boldsymbol{r}$ and $\boldsymbol{r}'$ are the position vectors of the same point in the two sets of coordinates and the coefficients a_{ij} form the matrix of the representation $D^{(1/2)}$ of the rotation group (see (12.34)). A two-component quantity of this kind, defined at each point in three-dimensional space, is a spinor field.

The probability that the electron will be found in a volume element dv at the end-point of the vector $\boldsymbol{r}$ is now defined by

$$(|\chi_1(\boldsymbol{r})|^2 + |\chi_2(\boldsymbol{r})|^2)\,dv \tag{13.18}$$

Since the matrix $\|\alpha_{ik}\|$ is unitary, the quantity given by (13.18) is invariant under the transformation (13.17):

$$|\chi_1'(\boldsymbol{r}')|^2 + |\chi_2'(\boldsymbol{r}')|^2 = |\chi_1(\boldsymbol{r})|^2 + |\chi_2(\boldsymbol{r})|^2 \qquad (13.19)$$

Let us now find the form of the infinitesimal operators for the spinor field. The transformation law given by (13.17) can be written in the form

$$\chi_i'(\boldsymbol{r}) = \sum_j \alpha_{ij}\chi_j(g^{-1}\boldsymbol{r}) \qquad (13.20)$$

or

$$\chi(\boldsymbol{r}) = D^{(1/2)}\chi(g^{-1}\boldsymbol{r})$$

Consider, for example, rotation about the x_3-axis. In this case, if we retain linear terms in α_3 we can write

$$\chi'(\boldsymbol{r}) = (\hat{E} + \hat{A}_3^{(1/2)}\alpha_3)\chi + \alpha_3\left(x_2\frac{\partial}{\partial x_1} - x_1\frac{\partial}{\partial x_2}\right)\chi \qquad (13.21)$$

where $\hat{A}_3^{(1/2)}$ is the infinitesimal operator of the irreducible representation $D^{(1/2)}$ of the rotation group. Hence it follows that the infinitesimal operator $\hat{A}_3$ for the spinor field is given by

$$\hat{A}_3 = \hat{A}_3^{(1/2)} + x_2\frac{\partial}{\partial x_1} - x_1\frac{\partial}{\partial x_2} \qquad (13.22a)$$

Similarly,

$$\hat{A}_1 = \hat{A}_1^{(1/2)} + x_3\frac{\partial}{\partial x_2} - x_2\frac{\partial}{\partial x_3} \qquad (13.22b)$$

$$\hat{A}_2 = \hat{A}_2^{(1/2)} + x_1\frac{\partial}{\partial x_3} - x_3\frac{\partial}{\partial x_1} \qquad (13.22c)$$

Let us now suppose that the Hamiltonian $\hat{H}$ for the electron is spherically symmetric. Invariance under the rotation group is described by

$$\hat{A}_i\hat{H} - \hat{H}\hat{A}_i = 0 \qquad (13.23)$$

where the $\hat{A}_i$ are defined by (13.22). Let us introduce the Hermitian operators

$$\hat{J}_k = i\hbar\hat{A}_k \quad (k = 1, 2, 3) \qquad (13.24)$$

In view of (13.23), these operators are constants of motion and are the total angular momentum operators. The operator $\hat{J}$ consists of two terms, one of which is the orbital angular momentum operator defined earlier $\hat{L}$. and the other $\hat{S} = i\hbar\hat{A}^{(1/2)}$ is the spin angular momentum. According to (11.28), in the representation in which $\hat{S}_3$ is diagonal, the operator $\hat{S}$ can be written in the form

$$\left.\begin{aligned} \hat{S}_1 &= \frac{\hbar}{2}\begin{pmatrix} 0 & 1 \\ 1 & 0 \end{pmatrix} = \frac{\hbar}{2}\sigma_1 \\ \hat{S}_2 &= \frac{\hbar}{2}\begin{pmatrix} 0 & i \\ -i & 0 \end{pmatrix} = \frac{\hbar}{2}\sigma_2 \\ \hat{S}_3 &= \frac{\hbar}{2}\begin{pmatrix} 1 & 0 \\ 0 & -1 \end{pmatrix} = \frac{\hbar}{2}\sigma_3 \end{aligned}\right\} \tag{13.25}$$

The matrices σ_1, σ_2 and σ_3 are the Pauli matrices. By applying $\hat{S}_3$ to the spinor χ, we obtain

$$\hat{S}_3\chi = \hbar\begin{pmatrix} \frac{1}{2} & 0 \\ 0 & -\frac{1}{2} \end{pmatrix}\begin{pmatrix} \chi_1 \\ \chi_2 \end{pmatrix} = \hbar\begin{pmatrix} \frac{1}{2}\chi_1 \\ -\frac{1}{2}\chi_2 \end{pmatrix} \tag{13.26}$$

Hence, it follows that the spinors $\begin{pmatrix} \chi_1 \\ 0 \end{pmatrix}$ and $\begin{pmatrix} 0 \\ \chi_2 \end{pmatrix}$ are the eigenvectors of $\hat{S}_3$ with eigenvalues $\frac{\hbar}{2}$ and $-\frac{\hbar}{2}$. Since an arbitrary state of the electron can be written as the superposition of such states, i.e.

$$\begin{pmatrix} \chi_1 \\ \chi_2 \end{pmatrix} = \begin{pmatrix} \chi_1 \\ 0 \end{pmatrix} + \begin{pmatrix} 0 \\ \chi_2 \end{pmatrix} \tag{13.27}$$

the quantity $|\chi_1|^2 dv$ gives the probability that the z-component of the spin angular momentum of an electron located in the volume element dv is equal to $\frac{\hbar}{2}$, while the quantity $|\chi_2|^2 dv$ has the same significance for the z-component $-\frac{\hbar}{2}$.

Consider the approximation corresponding to (13.1), when the spin operators can be neglected in the Hamiltonian. The electron wave function can then be written in the form

$$\begin{pmatrix} \psi_1(\boldsymbol{r}) \\ \psi_2(\boldsymbol{r}) \end{pmatrix} = \varphi(\boldsymbol{r}) \begin{pmatrix} \chi_1 \\ \chi_2 \end{pmatrix} \tag{13.28}$$

where $\begin{pmatrix} \chi_1 \\ \chi_2 \end{pmatrix}$ is a constant spinor and the function $\varphi(\boldsymbol{r})$ is a solution of the Schroedinger equation. The indices of the spinor components can be conveniently taken to be the corresponding eigenvalues of the operator $i\hat{A}_3 = \frac{1}{\hbar}\hat{S}_3$, i.e. $\frac{1}{2}$ and $-\frac{1}{2}$. Occasionally, they are regarded as the arguments of a function which can assume only two values:

$$\psi(\boldsymbol{r}, \sigma) = \varphi(\boldsymbol{r})\chi(\sigma) = \begin{cases} \varphi(\boldsymbol{r})\chi_{1/2}, & \sigma = \frac{1}{2} \\ \varphi(\boldsymbol{r})\chi_{-1/2}, & \sigma = -\frac{1}{2} \end{cases} \tag{13.29}$$

If the Schroedinger equation for our problem is invariant under the rotation group, this eigenfunction should transform by irreducible representations of the rotation group $D^{(l)}$. If we then take the spin state into account, we find that each energy level is $2(2l+1)$-fold degenerate and the corresponding wave functions can be chosen to be

$$\varphi_{lm}(\boldsymbol{r})\,\delta_{\sigma,\,1/2}, \quad \varphi_{lm}\delta_{\sigma,\,-1/2}$$

or

$$\begin{pmatrix} \varphi_{lm} \\ 0 \end{pmatrix}, \quad \begin{pmatrix} 0 \\ \varphi_{lm} \end{pmatrix} \qquad (m = -l, -l+1, \ldots, l) \tag{13.30}$$

It is clear that these functions transform by an irreducible representation which is the direct product $D^{(l)} \times D^{(1/2)}$. By the Clebsch-Gordan rule we have

$$D^{(l)} \times D^{(1/2)} = D^{(l-1/2)} \oplus D^{(l+1/2)} \tag{13.31}$$

From the functions (13.30) we can construct new functions which transform by irreducible representations or, in other words, we can find the eigenfunctions of the operators $\hat{J}_2$ and $\hat{J}_3$. The final result is

$$\Psi_{JM} = c^{(l,\,1/2,\,J)}_{M-1/2,\,1/2,\,M}\begin{pmatrix}\varphi_{l,\,M-1/2}(\boldsymbol{r})\\ 0\end{pmatrix} + c^{(l,\,1/2,\,J)}_{M+1/2,\,-1/2,\,M}\begin{pmatrix}0\\ \varphi_{l,\,M+1/2}(\boldsymbol{r})\end{pmatrix}$$

$$= \begin{pmatrix} c^{(l,\,1/2,\,J)}_{M-1/2,\,1/2,\,M}\varphi_{l,\,M-1/2}(\boldsymbol{r}) \\ c^{(l,\,1/2,\,J)}_{M+1/2,\,-1/2,\,M}\varphi_{l,\,M+1/2}(\boldsymbol{r}) \end{pmatrix} \tag{13.32}$$

13.4 Kramers' theorem

If the Hamiltonian of an electron commutes with the operator

$$\hat{\Theta} = \frac{2}{\hbar}\hat{S}_2 K \tag{13.33}$$

where K represents the operation of complex conjugation and $\hat{S}_2$ is the spin angular momentum operator along the x_2-axis, then the energy levels of the electron are twofold degenerate. (Kramers' theorem has been proved for a system consisting of an odd number of electrons. In this case, $\hat{\Theta} = \left(\frac{2}{\hbar}\hat{S}_2^{(1)}\right)\dots\left(\frac{2}{\hbar}\hat{S}_2^{(n)}\right)K$, where $\hat{S}_2^{(i)}$ refers to the i-th electron.) If the system under consideration exhibits additional symmetry and the associated degeneracy, the resultant degree of degeneracy must be even.

We shall show, to begin with, that the Hamiltonian commutes with $\hat{\Theta}$ provided only that it does not contain a term describing the interaction with a magnetic field. In fact, in this case, we need only add the spin-orbital interaction term to the Hamiltonian $\hat{H}_0$ of the Schroedinger equation (13.1), which is of the form

$$\hat{H}_{SO} = \frac{1}{2m^2c^2}\left(\frac{1}{r}\frac{dV}{dr}\right)(\hat{\boldsymbol{L}},\,\hat{\boldsymbol{S}}) \tag{13.34}$$

We note that the Hamiltonian $\hat{H}_0$ and the operator (13.34) have been written for a spherically symmetric potential. This was done for the sake of simplicity but it is not important for the ensuing analysis.

Since the Hamiltonian $\hat{H}_0$ is real and does not by

hypothesis contain spin operators, it is clear that

$$\hat{\Theta}\hat{H}_0 = \hat{H}_0\hat{\Theta} \tag{13.35}$$

We shall now show that

$$\hat{\Theta}\hat{H}_{SO} = \hat{H}_{SO}\hat{\Theta} \tag{13.36}$$

In fact, according to (13.10) the operator $\hat{\boldsymbol{L}}$ is purely imaginary and, therefore,

$$\hat{\Theta}\hat{\boldsymbol{L}} = -\hat{\boldsymbol{L}}\hat{\Theta} \tag{13.37}$$

Using the explicit form of the matrix $\hat{S}_i$, we can readily show that

$$\hat{\Theta}\hat{\boldsymbol{S}} = -\hat{\boldsymbol{S}}\hat{\Theta} \tag{13.38}$$

Equation (13.36) then follows from (13.37) and (13.38). If we now place our system in a magnetic field $\boldsymbol{H}$, the Hamiltonian will contain the additional term

$$-\frac{e}{2mc}\boldsymbol{H}(\boldsymbol{L}+2\boldsymbol{S}) \tag{13.39}$$

which by (13.37) and (13.38) does not commute with $\hat{\Theta}$.

In view of (13.37) and (13.38) and the relations

$$\hat{\Theta}\hat{\boldsymbol{p}} = -\hat{\boldsymbol{p}}\hat{\Theta}, \quad \hat{\Theta}\boldsymbol{r} = \boldsymbol{r}\hat{\Theta} \tag{13.40}$$

which are readily verified, the operator $\hat{\Theta}$ is usually referred to as the time reversal operator.

Let us now go on to the proof of Kramers' theorem. It is readily verified that

$$\hat{\Theta}^2 = -1 \tag{13.41}$$

Consider two electron states described by the spinors

$$\boldsymbol{\Psi}(\boldsymbol{r}) = \begin{pmatrix}\psi_1(\boldsymbol{r})\\ \psi_2(\boldsymbol{r})\end{pmatrix}, \quad \boldsymbol{\Phi}(\boldsymbol{r}) = \begin{pmatrix}\varphi_1(\boldsymbol{r})\\ \varphi_2(\boldsymbol{r})\end{pmatrix} \tag{13.42}$$

We shall define their scalar product by

$$(\boldsymbol{\Psi}, \boldsymbol{\Phi}) = \psi_1\bar{\varphi}_1 + \psi_2\bar{\varphi}_2 \tag{13.43}$$

so that, using the fact that $\hat{S}_2$ is Hermitian, we may write

$$(\hat{\Theta}\boldsymbol{\Psi}, \hat{\Theta}\boldsymbol{\Phi}) = \overline{(\boldsymbol{\Psi}, \boldsymbol{\Phi})} = (\boldsymbol{\Phi}, \boldsymbol{\Psi}) \tag{13.44}$$

and hence

$$(\Psi, \hat{\Theta}\Psi) = \overline{(\hat{\Theta}\Psi, \hat{\Theta}^2\Psi)} = (\hat{\Theta}^2\Psi, \hat{\Theta}\Psi) = -(\Psi, \hat{\Theta}\Psi) \quad (13.45)$$

so that

$$(\Psi, \hat{\Theta}\Psi) = 0 \quad (13.46)$$

From the last equation we may conclude that the states Ψ and $\hat{\Theta}\Psi$ are orthogonal and, therefore, linearly independent. On the other hand, by (13.33) these states belong to the same energy level, and this proves the theorem.

We must make one further important remark. If the system under consideration is spatially symmetric, the degeneracy introduced by it may include the Kramers degeneracy. Invariance under time reversal will not then lead to additional degeneracy. As an example, consider the case where the symmetry group of the system coincides with the rotation group $O^+(3)$. We note that, in this case, the state of an electron is characterized by the resultant angular momentum J, and the energy eigenvalue is $(2J+1)$-fold degenerate with respect to the eigenvalue M of the z-component of this angular momentum. Using (13.32) we readily find a more general form of the wave function

$$\Psi_{JM}(r) = R_{J-\frac{1}{2}}(r) \begin{Vmatrix} C^{\left(J-\frac{1}{2},\, \frac{1}{2},\, J\right)}_{M-\frac{1}{2},\, \frac{1}{2},\, M} Y^{\left(M-\frac{1}{2}\right)}_{J-\frac{1}{2}}(\theta, \varphi) \\ C^{\left(J-\frac{1}{2},\, \frac{1}{2},\, J\right)}_{M+\frac{1}{2},\, -\frac{1}{2},\, M} Y^{\left(M+\frac{1}{2}\right)}_{J-\frac{1}{2}}(\theta, \varphi) \end{Vmatrix}$$

$$+ R_{J+\frac{1}{2}}(r) \begin{Vmatrix} C^{\left(J+\frac{1}{2},\, \frac{1}{2},\, J\right)}_{M-\frac{1}{2},\, \frac{1}{2},\, M} Y^{\left(M-\frac{1}{2}\right)}_{J+\frac{1}{2}}(\theta, \varphi) \\ C^{\left(J+\frac{1}{2},\, \frac{1}{2},\, J\right)}_{M+\frac{1}{2},\, -\frac{1}{2},\, M} Y^{\left(M+\frac{1}{2}\right)}_{J+\frac{1}{2}}(\theta, \varphi) \end{Vmatrix} \quad (13.47)$$

$$(M = -J, \ -J+1, \ \ldots, \ J)$$

where $R_{J-\frac{1}{2}}(r)$ and $R_{J+\frac{1}{2}}(r)$ are certain real functions of $|\mathbf{r}|$.

We recall that $Y_l^{(m)} = P_l^{(m)} e^{im\varphi}$, $P_l^{(-m)} = (-1)^m P_l$. The Clebsch-Gordan coefficients are also real and have a number of properties of which

$$c^{(l_1 l_2 J)}_{m_1 m_2 M} = c^{(l_1 l_2 J)}_{-m_1, -m_2, -M} \tag{13.48}$$

will be useful in our analysis. Let us apply the operator $\hat{\Theta}$ to the function $\boldsymbol{\Psi}_{JM}(\boldsymbol{r})$. In our case,

$$\hat{\Theta} = \begin{pmatrix} 0 & i \\ -i & 0 \end{pmatrix} K \tag{13.49}$$

We shall confine our attention to the first component. The transformation of the second component is entirely analogous Using the definition of spherical harmonics and the property given by (13.48), we have

$$\begin{aligned} \hat{\Theta} R_{J-\frac{1}{2}}(r) & \begin{bmatrix} c^{(J-\frac{1}{2},\, \frac{1}{2},\, J)}_{M-\frac{1}{2},\, \frac{1}{2},\, M} Y^{(M-\frac{1}{2})}_{J-\frac{1}{2}}(\theta, \varphi) \\ c^{(J-\frac{1}{2},\, \frac{1}{2},\, J)}_{M+\frac{1}{2},\, -\frac{1}{2},\, M} Y^{(M+\frac{1}{2})}_{J-\frac{1}{2}}(\theta, \varphi) \end{bmatrix} \\ = R_{J\mp\frac{1}{2}}(r) & \begin{bmatrix} i c^{(J-\frac{1}{2},\, \frac{1}{2},\, J)}_{M+\frac{1}{2},\, -\frac{1}{2},\, M} \overline{Y}^{(M+\frac{1}{2})}_{J-\frac{1}{2}}(\theta, \varphi) \\ -i c^{(J-\frac{1}{2},\, \frac{1}{2},\, J)}_{M-\frac{1}{2},\, \frac{1}{2},\, M} \overline{Y}^{(M-\frac{1}{2})}_{J-\frac{1}{2}}(\theta, \varphi) \end{bmatrix} \\ = R_{J-\frac{1}{2}}(r)\, i(-1)^{M+\frac{1}{2}} & \begin{bmatrix} c^{(J-\frac{1}{2},\, \frac{1}{2},\, J)}_{-(M+\frac{1}{2}),\, \frac{1}{2},\, -M} Y^{(-(M+\frac{1}{2}))}_{J-\frac{1}{2}}(\theta, \varphi) \\ c^{(J,\, -\frac{1}{2},\, \frac{1}{2})}_{-(M-\frac{1}{2}),\, -\frac{1}{2},\, -M} Y^{(-(M-\frac{1}{2}))}_{J-\frac{1}{2}}(\theta, \varphi) \end{bmatrix} \end{aligned} \tag{13.50}$$

Thus

$$\hat{\Theta} \boldsymbol{\Psi}_{JM}(\boldsymbol{r}) = i(-1)^{M+\frac{1}{2}} \boldsymbol{\Psi}_J^{-M}(\boldsymbol{r}) \tag{13.51}$$

It is therefore evident that the operator $\hat{\Theta}$ connects states with opposite values of components of the total angular momentum and, in the present case, does not lead to additional degeneracy.

Chapter 14

Additional Degeneracy in a Spherically Symmetric Field

In Chapter 13 we investigated the classification of the eigenfunctions of the Schroedinger equation for an arbitrary spherically symmetric potential. There are, however, two types of spherically symmetric potential for which there is additional degeneracy and, consequently, the above classification is incomplete. These two potentials are the Coulomb potential $U = \frac{1}{r}$ and the harmonic oscillator potential $U = kr^2$. We shall show in this chapter that the additional degeneracy is connected with the invariance of the corresponding Schroedinger equation under certain definite transformation groups, of which the rotation group is merely the subgroup.

14.1 Additional degeneracy

The existence of additional degeneracy for the above potentials is known from the solutions of the corresponding Schroedinger equations.

Consider the Schroedinger equation with the Coulomb potential, i.e. the equation for the hydrogen atom. In atomic units ($e = 1$, $m = 1$, $\hbar = 1$) this equation is of the form

$$\left(-\frac{1}{2}\Delta - \frac{1}{r}\right)\psi(r) = E\psi(r) \tag{14.1}$$

It is well known that the eigenvalues of this equation which correspond to bound states are governed by the principal quantum number n and are given by

$$E_n = -\frac{1}{2}\frac{1}{n^2} \tag{14.2}$$

For given n, the azimuthal quantum number can assume the values $l = n-1,\ n-2,\ \ldots,\ 1,\ 0$. This means that the wave functions corresponding to a degenerate level E_n transform in accordance with a reducible representation of the rotation group, which can be decomposed into irreducible representations $D^{(n-1)},\ D^{(n-2)},\ \ldots,\ D^{(0)}$.

Additional degeneracy will also occur for the energy levels of the isotropic harmonic oscillator, for which the Schroedinger equation (in atomic units) is

$$\left(-\frac{1}{2\mu}\Delta + \frac{\mu\omega^2}{2}r^2\right)\psi(r) = E\psi(r) \tag{14.3}$$

To simplify the notation let us set $\mu = \omega = 1$. We shall seek the solution of (14.3) in the form

$$\varphi(r) = Y_{lm}(\theta, \varphi)\frac{1}{r}f(r)e^{-r^2/2}, \quad f(0) = 0 \tag{14.4}$$

For the function $f(r)$ we have

$$f'' - 2rf' + \left[2n + 2 - \frac{l(l+1)}{r^2}\right]f = 0 \tag{14.5}$$

where

$$n = E - \frac{3}{2} \tag{14.6}$$

The solution of this equation, which satisfies the boundary condition at the origin, can be expressed in terms of the confluent hypergeometric function

$$f(r) = r^{l+1}F\left(-\frac{n-l}{2},\ l + \frac{3}{2},\ r^2\right) \tag{14.7}$$

To ensure that the wave function tends to zero as $r \to \infty$, the series representing the hypergeometric function must stop at a certain term. This is only possible if $\frac{n-l}{2}$ is an integer or zero. and hence it follows that n must be an integer. For fixed n which determines the energy eigenvalue, the azimuthal number l can assume the values $n, n-2, n-4, \ldots$. The corresponding wave function will therefore transform in accordance with a reducible representation of the rotation group, which decomposes into the irreducible representations $D^{(n)}, D^{(n-2)}, \ldots$

14.2 Connection with classical mechanics

The exceptional properties of the above potentials can also be seen in classical mechanics. We recall that, classically, the motion of a particle in a central-force potential $U(r)$ takes place in the plane perpendicular to the direction of the angular momentum $\boldsymbol{L}=[\boldsymbol{r}, \boldsymbol{p}]$. If we introduce polar coordinates in this plane with the origin at the centre of force, the polar angle φ and the position vector $\boldsymbol{r}$ of the particle will be related by

$$\varphi = \int \frac{L\,dr}{r^2\sqrt{2\mu\,(E-U(r))-\frac{L^2}{r^2}}} + \text{const} \tag{14.8}$$

where E is the energy and μ the mass of the particle.

The inequality

$$U(r)+\frac{L^2}{2\mu r^2}<E \tag{14.9}$$

defines the domain of the position vector $\boldsymbol{r}$. This region has, in any case, the lower bound $r=r_{\min}$, since the left-hand side of the inequality contains the term $\frac{L}{2\mu r^2}$, which increases as $r\to 0$. If, moreover, there is an upper bound $r=r_{\max}$, the motion is called finite and the corresponding trajectory lies entirely within a ring with outer and inner radii equal to $r_{\max}$ and $r_{\min}$, respectively. For an arbitrary central field, the trajectory touches the inner and outer circles of this ring an infinite number of times, and is not closed (Fig. 16) .

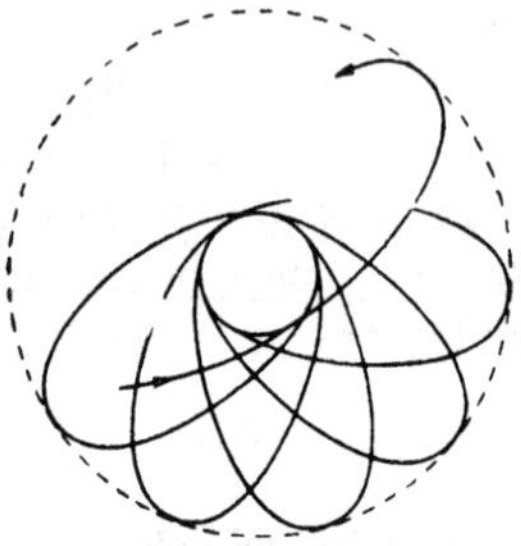

Fig. 16

This occurs because the quantity

$$\Delta\varphi = \int_{r_{min}}^{r_{max}} \frac{L\,dr}{r^2\sqrt{2\mu\,[E - U(r)] - \frac{L^2}{r^2}}}$$

is not, in general, a rational part of 2π. For the Coulomb and the harmonic oscillator potentials, however, the trajectories will be closed, since for the Coulomb field $\Delta\varphi = \frac{\pi}{2}$ and for the isotropic oscillator $\Delta\varphi = \frac{\pi}{4}$ for any E and L.

14.3 Symmetry group of the hydrogen atom

The symmetry group of the Schroedinger equation for the hydrogen atom was investigated by Fock in 1935. (V. A. Fock, Izv. AN SSSR, OTN, pp.169-179 (1935)). Transformations in this group cannot be reduced to transformations in ordinary three-dimensional space. Moreover, the symmetry groups of this equation for bound ($E < 0$) and free ($E > 0$) states are different. In this section we shall consider the case of the discrete spectrum in greater detail.

The Schroedinger equation (14.1) for the hydrogen atom in the momentum representation is

$$\frac{1}{2}p^2\varphi(\boldsymbol{p})+\frac{1}{2\pi}\int\frac{\varphi(\boldsymbol{p}')\,d\boldsymbol{p}'}{(\boldsymbol{p}-\boldsymbol{p}')^2}=-\frac{p_0^2}{2}\varphi(\boldsymbol{p}) \tag{14.10}$$

where

$$\varphi(\boldsymbol{p})=(2\pi)^{-3/2}\int\psi(\boldsymbol{r})\,e^{-i(\boldsymbol{pr})}d\boldsymbol{r},\quad -\frac{p_0^2}{2}=E \tag{14.11}$$

Let us map each point $\boldsymbol{p}$ in momentum space on a point on a unit sphere in four-dimensional space with the coordinates ξ, η, ζ, χ:

$$\left.\begin{aligned}\xi&=\frac{2p_0p_x}{p_0^2+p^2}\quad \zeta=\frac{2p_0p_z}{p_0^2+p^2}\\ \eta&=\frac{2p_0p_y}{p_0^2+p^2}\quad \chi=\frac{p_0^2-p^2}{p_0^2+p^2}\\ &\xi^2+\eta^2+\zeta^2+\chi^2=1\end{aligned}\right\} \tag{14.12}$$

If we now introduce the function

$$\Phi(\boldsymbol{p})=\frac{\pi}{\sqrt{8}}p_0^{-5/2}(p_0^2+p^2)^2\varphi(\boldsymbol{p})=\Psi(\xi, \eta, \zeta, \chi) \tag{14.13}$$

then Equation (14.11) can be transformed to read

$$\Psi(\xi, \eta, \zeta, \chi)=\frac{1}{2\pi^2p_0}\int\frac{\Psi(\xi', \eta', \zeta', \chi')\,d\Omega}{|\xi-\xi'|^2+|\eta-\eta'|^2+|\zeta-\zeta'|^2+|\chi-\chi'|^2} \tag{14.14}$$

where $d\Omega$ is an element of area on the four-dimensional sphere of unit radius over which the integral is evaluated. This equation is invariant under rotations in four-dimensional space and, consequently, $O^+(4)$ is its symmetry group. On the basis of Wigner's theorem we can then conclude that the eigenfunctions of this equation should transform by the irreducible representations of the group $O^+(4)$. The irreducible representations of this group will be derived in Chapter 21. We can now verify that the classification of states based on these representations of the group $O^+(4)$ is the same as the usual classification of the states of the hydrogen atom, and explains the additional degeneracy.

It is interesting to obtain the explicit expressions for

the infinitesimal operators of representations of the four-dimensional rotation group (Yu. A. Dobronravov, Vest. Leningr. Gos. Univ., No.10, p. 5 (1957)).

In our problem they are the quantum-mechanical constants of motion.

In four-dimensional Euclidean space we can consider the six independent two-dimensional planes $\xi\eta$, $\xi\zeta$, $\xi\chi$, $\eta\zeta$, $\eta\chi$, $\zeta\chi$. Therefore, we can take the rotation parameters in these two-dimensional planes as the six parameters on which an arbitrary real orthogonal matrix of order four will depend. Infinitesimal rotations in these planes correspond to the transformations

$$\left.\begin{matrix} \begin{cases} \xi' = \xi - \alpha_3\eta \\ \eta' = \alpha_3\xi + \eta \end{cases} & \begin{cases} \xi' = \xi + \alpha_2\zeta \\ \zeta' = -\alpha_2\xi + \zeta \end{cases} & \begin{cases} \eta' = \eta - \alpha_1\zeta \\ \zeta' = \alpha_1\eta + \zeta \end{cases} \\ \begin{cases} \xi' = \xi + \beta_1\chi \\ \chi' = -\beta_1\xi + \chi \end{cases} & \begin{cases} \eta' = \eta + \beta_2\chi \\ \chi' = -\beta_2\eta + \chi \end{cases} & \begin{cases} \zeta' = \zeta + \beta_3\chi \\ \chi' = -\beta_3\zeta + \chi \end{cases} \end{matrix}\right\} \tag{14.15}$$

According to (14.12), these rotations correspond in the momentum space to the transformations

$$\left.\begin{matrix} \begin{cases} p'_y = p_y - \alpha_1 p_z \\ p'_z = \alpha_1 p_y + p_z \end{cases} \quad \begin{cases} p'_x = p_x + \alpha_2 p_z \\ p'_z = -\alpha_2 p_x + p_z \end{cases} \quad \begin{cases} p'_x = p_x - \alpha_3 p_y \\ p'_y = \alpha_3 p_x + p_y \end{cases} \\ \begin{cases} p'_x = p_x + \dfrac{2p_x^2 + p_0^2 - p^2}{2p_0}\beta_1 \\ p'_y = p_y + \dfrac{p_y p_x}{p_0}\beta_1 \\ p'_z = p_z + \dfrac{p_z p_x}{p_0}\beta_1 \end{cases} \quad \begin{cases} p'_x = p_x + \dfrac{p_x p_y}{p_0}\beta_2 \\ p'_y = p_y + \dfrac{2p_y^2 + p_0^2 - p^2}{2p_0}\beta_2 \\ p'_z = p_z + \dfrac{p_z p_y}{p_0}\beta_2 \end{cases} \\ \begin{cases} p'_x = p_x + \dfrac{p_x p_z}{p_0}\beta_3 \\ p'_y = p_y + \dfrac{p_y p_z}{p_0}\beta_3 \\ p'_z = p_z + \dfrac{2p_z^2 + p_0^2 - p^2}{2p_0}\beta_3 \end{cases} \end{matrix}\right\} \tag{14.16}$$

The latter transformations in their turn induce infinitesimal transformations in the space of the functions

$$\delta\Phi = \frac{\partial\Phi}{\partial p_x}\,\delta p_x + \frac{\partial\Phi}{\partial p_y}\,\delta p_y + \frac{\partial\Phi}{\partial p_z}\,\delta p_z \tag{14.17}$$

Bearing in mind the relation between the functions $\Phi(\boldsymbol{p})$ and $\varphi(\boldsymbol{p})$, we have

$$\delta_{\alpha_1}\varphi = \alpha_1\left(-\,p_z\frac{\partial\varphi}{\partial p_y} + p_y\frac{\partial\varphi}{\partial p_z}\right) \tag{14.18}$$

$$\delta_{\beta_1}\varphi = \beta_1\left[\frac{p_x^2 - p_y^2 - p_z^2 + p_0^2}{2p_0}\,\frac{\partial\varphi(\boldsymbol{p})}{\partial p_x}\right.$$

$$\left. + \frac{p_x p_y}{p_0}\,\frac{\partial\varphi(\boldsymbol{p})}{\partial p_y} + \frac{p_x p_z}{p_0}\,\frac{\partial\varphi(\boldsymbol{p})}{\partial p_z} + 2\,\frac{p_x}{p_0}\,\varphi(\boldsymbol{p})\right] \tag{14.19}$$

and four further equations which are obtained by cyclic permutation of x, y, z.

We have just written the infinitesimal operators corresponding to the infinitesimal transformations (14.15) in the space of the wave functions $\varphi(\boldsymbol{p})$ corresponding to the level $E = -\frac{p_0^2}{2}$. To obtain infinitesimal operators of the four-dimensional rotation group, acting on any function which can be expanded in terms of the eigenfunctions corresponding to the discrete spectrum, we must replace p_0 by the operator $\sqrt{-2\hat{H}}$. Moreover, if we replace the derivatives $\frac{\partial}{\partial p_x}, \frac{\partial}{\partial p_y}, \frac{\partial}{\partial p_z}$ by ix, iy, iz, respectively, we obtain the following final expression for the six infinitesimal operators

$$\left.\begin{aligned} \hat{L}_x &= \hat{p}_z y - \hat{p}_y z \\ \hat{L}_y &= \hat{p}_x z - \hat{p}_z x \\ \hat{L}_z &= \hat{p}_y x - \hat{p}_x y \end{aligned}\right\} \tag{14.20}$$

$$\left.\begin{aligned} \hat{N}_x &= \frac{1}{2}\left[\frac{[\hat{p}, \hat{L}]_x - [\hat{L}, \hat{p}]_x}{2} - \frac{x}{r}\right]\sqrt{\frac{2}{-\hat{H}}} \\ \hat{N}_y &= \frac{1}{2}\left[\frac{[\hat{p}, \hat{L}]_y - [\hat{L}, \hat{p}]_y}{2} - \frac{y}{r}\right]\sqrt{\frac{2}{-\hat{H}}} \\ \hat{N}_z &= \frac{1}{2}\left[\frac{[p, \hat{L}]_z - [\hat{L}, \hat{p}]_z}{2} - \frac{z}{r}\right]\sqrt{\frac{2}{-\hat{H}}} \end{aligned}\right\} \tag{14.21}$$

The first three are the ordinary components of the angular momentum operator. They exist for all spherically symmetric potentials. The operators (14.21) are characteristic of the Coulomb field only. They can be written in the vector form

$$\hat{N} = \frac{1}{2}\left[\frac{[\hat{p}, \hat{L}] - [\hat{L}, \hat{p}]}{2} - \frac{r}{r}\right]\sqrt{\frac{2}{-\hat{H}}} \qquad (14.22)$$

The corresponding classical quantity is the position vector of the second focus of the ellipse. Thus this constant of motion fixes the direction of the perihelion of the orbit, and this is connected with the fact that the orbit is closed.

V. A. Fock has shown that for states in the continuous spectrum the equation given by (14.1) is invariant under a transformation group which is isomorphic to the Lorentz group.

14.4 Symmetry group of the harmonic oscillator

Having established the existence of additional degeneracy for the states of the hydrogen atom, let us now determine the complete symmetry group of the isotropic oscillator. This group was found independently by Demkov, Hill, and Jauch. (Yu. N. Demkov, Vest. Leningr. Gos. Univ., No.11 (1953); F. L. Hill and J. M. Jauch, Phys. Rev., V.57, p.641 (1950). Here we shall follow the treatment given by Baker, Phys. Rev., V. 103, p.1119 (1956).)

The three-dimensional oscillator is a special case of the n-dimensional isotropic oscillator. Since our analysis of the symmetry group will not depend on the dimensionality of the oscillator, we may just as well consider the general case.

The Hamiltonian for the isotropic n-dimensional oscillator (in atomic units) is

$$H = \sum_{k=1}^{n}\left(\frac{p_k^2}{2} + \frac{\omega^2}{2} q_k^2\right) \qquad (14.23)$$

We are assuming, for simplicity, that the mass $m = 1$.

Consider the operators

$$\left.\begin{aligned} a_k &= (2\omega)^{-1/2}(\omega q_k + ip_k) \\ a_k^+ &= (2\omega)^{-1/2}(\omega q_k - ip_k) \end{aligned}\right\} \tag{14.24}$$

From the commutation relations for p_k and q_k we can readily obtain the commutation relation for the operators a_k and a_k^+:

$$a_k a_l - a_l a_k = 0, \quad a_k^+ a_l^+ - a_l^+ a_k^+ = 0, \quad a_k a_l^+ - a_l^+ a_k = \delta_{kl} \tag{14.25}$$

The Hamiltonian (14.23) can be written in the form

$$H = \omega \sum_{k=1}^{n} \left(a_k^+ a_k + \frac{1}{2} \right) \tag{14.26}$$

We can now replace the operators a_k by the new operators a_k' which are related to the old operators by an arbitrary unitary transformation:

$$a_k' = \sum_s u_{ks} a_s, \quad a_k^{+\prime} = \sum_s u_{sk}^+ a_s^+ \tag{14.27}$$

It is readily verified that since the matrix $\| u_{ik} \|$ is unitary, the operators a_k' and $a_l'^+$ satisfy the original commutation relations.

In fact, we have

$$\left.\begin{aligned} a_k' a_l' - a_l' a_k' &= \sum_{i,j} u_{ki} u_{lj} (a_i a_j - a_j a_i) = 0 \\ a_k' a_l'^+ - a_l'^+ a_k' &= \sum_{i,j} u_{kj} u_{il}^+ (a_j a_i^+ - a_i^+ a_j) = \\ &= \sum_{i,j} u_{kj} u_{jl}^+ = \delta_{kl} \end{aligned}\right\} \tag{14.28}$$

It is also clear that the Hamiltonian (14.26) should be invariant under the transformation (14.27). In fact,

$$a_k = \sum_s u_{ks}^+ a_s', \quad a_k^+ = \sum_r u_{rk}^+ a_r^{+\prime} \tag{14.29}$$

$$\begin{aligned} H &= \omega \sum_k \left(a_k^+ a_k + \frac{1}{2} \right) = \omega \sum_k \left\{ \sum_{s,r} \left(a_r'^+ a_s' u_{ks}^+ u_{rk} + \frac{1}{2} \right) \right\} \\ &= \omega \sum_{s,r} \left(a_r'^+ a_s' \delta_{rs} + \frac{1}{2} \right) = \omega \sum_r \left(a_r'^+ a_r' + \frac{1}{2} \right) \end{aligned} \tag{14.30}$$

Hence it follows that the symmetry group of the n-dimensional isotropic oscillator is the n-dimensional unitary group $U(n)$

For the three-dimensional oscillator, the symmetry group is therefore the three-dimensional unitary group $U(3)$ which contains the rotation group as a real sub-group. Consequently, if we consider the matrix $\|u_{ik}\|$ as real, the transformations (14.27) can be written in the form

$$a'_k = \sum_{s=1}^{3} u_{ks} a_s, \quad a'^{+}_k = \sum_{s=1}^{3} u_{ks} a^{+}_s \tag{14.31}$$

This transformation of the operators a_s, a^{+}_s corresponds to the same transformations for the operators q_k and p_k:

$$q'_k = \sum_s u_{ks} q_s, \quad p'_k = \sum_s u_{ks} p_s \tag{14.32}$$

which are ordinary rotations in three-dimensional space.

Let us return now to the n-dimensional isotropic oscillator and find the representations by which its degenerate wave functions transform. The isotropic n-dimensional oscillator can be regarded as a set of n simple one-dimensional oscillators. A possible wave function, therefore, can be written as the product

$$\Phi_{m_1, \ldots, m_n} = \varphi_{m_1}(q_1)\varphi_{m_2}(q_2) \cdots \varphi_{m_n}(q_n) \tag{14.33}$$

where φ_{m_l} is the wave function of the one-dimensional oscillator with energy eigenvalue $\omega\left(m_l + \frac{1}{2}\right)$. The wave function (14.33) corresponds to the energy eigenvalue

$$E = \omega\left(m_1 + m_2 + \ldots + m_n + \frac{n}{2}\right) \tag{14.34}$$

It is clear that all the functions $\Phi_{m'_1, \ldots, m'_n}$ for which

$$\omega\left(m'_1 + m'_2 + \ldots + m'_n + \frac{n}{2}\right) = E \tag{14.35}$$

correspond to this energy. The ground state of our oscillator is described by the unique function

$$\Phi_{0\ldots 0} = \varphi_0(q_1) \cdots \varphi_0(q_n) \tag{14.36}$$

It can be shown that the function $\Phi_{0\ldots0}$ transforms by the identity representation of the unitary group. The wave functions for the excited states can be constructed by applying the operators a_k^+ to the ground-state functions. If we omit the normalizing factors we can write

$$\Phi_{m_1 \ldots m_n} = (a_1^+)^{m_1} (a_2^+)^{m_2} \ldots (a_n^+)^{m_n} \Phi_{0\ldots0} \tag{14.37}$$

and, consequently, the eigenfunctions corresponding to the eigenvalue $m + \frac{n}{2}$ can be written in the form

$$a_{i_1}^+ a_{i_2}^+ \ldots a_{i_m}^+ \Phi_{0\ldots0} \tag{14.38}$$

Since the order in which the operators a_i^+ are arranged is unimportant, these functions will transform under (14.27) as the components of a symmetric tensor of rank m in n-dimensional space.

In the theory of representations of the unitary group it is shown that representations realized on the components of a symmetric tensor are irreducible (see, for example, H. Weyl, Classical Groups). These representations do not, however, exhaust all the irreducible representations of the unitary group. Lack of space prevents us from analysing this problem in greater detail, and we shall merely calculate the order of the representation which can be realized on the components of a symmetric tensor of rank m in n-dimensional space. We know that the order of the representation is equal to the degree of degeneracy of the corresponding energy level.

Let us determine the number of independent components of a symmetric tensor of rank m in n-dimensional space. Let $\Phi_{j_1 \ldots j_m}$ be a component of this tensor where, in view of the symmetry of the tensor, we can always arrange the subscripts in a non-descending order:

$$j_1 \leqslant j_2 \leqslant \ldots \leqslant j_m$$

In that case we have

$$i_1 < i_2 < \ldots < i_m$$

where $l_k = j_k + k - 1$. The numbers l_k are all different and can assume values between 1 and $n+m-1$. The degree of degeneracy of the level $n+m-1$ of the isotropic m-dimensional oscillator is therefore equal to $E = m + \frac{n}{2}$

$$\frac{(n+m-1)!}{m!\,(n-1)!} \tag{14.39}$$

Since we have already digressed quite considerably, let us take our discussion a little further by briefly considering the symmetry properties of the wave functions of the n-dimensional isotropic oscillator whose coordinates transform under unitary and real representations of a symmetry group (for example, a point group), as in the case of the normal vibrations of molecules. This coordinate transformation will give rise to an analogous transformation of the operators a_k^+. Therefore, in view of the above discussion, we may conclude that the wave functions belonging to the energy eigenvalue $m + \frac{n}{2}$ will transform under an m-th order 'symmetric' power of the representation under which the normal coordinates are transformed (Chapter 16).

Chapter 15

Permutation Groups

In this chapter we shall consider the group S_n of permutations of n symbols and establish its irreducible representations. The specific realization of this group in quantum mechanics is the group of permutations of variables describing identical particles. In order to be clear about the aim of our discussion right from the beginning, let us start with the formulation of the quantum-mechanical problem.

15.1 Quantum-mechanical description of a system of identical particles

Consider the Schroedinger equation for a system consisting of n identical particles:

$$H(q_1, q_2, \ldots, q_n)\Psi(q_1, \ldots, q_n) = E\Psi(q_1, q_2, \ldots, q_n) \quad (15.1)$$

where q_i is the set of variables referring to the i-th particle. Since the Hamiltonian $H(q_1, q_2, \ldots, q_n)$ must be invariant under the interchange of the variables q_i, the solution of (15.1) should transform according to an irreducible representation of this group.

To ensure that the theory agrees with experiment it is necessary that the wave function for a system of particles with half-integral spins be antisymmetric under the interchange of any pair of particles, while the wave function for a system of particles with integral spins must be symmetric under this operation:

$$\left.\begin{aligned} &\Psi(q_1, q_2, \ldots, q_n) = -\Psi(q_2, q_1, \ldots, q_n) \quad \text{half-integral spin} \\ &\Psi(q_1, q_2, \ldots, q_n) = \Psi(q_2, q_1, \ldots, q_n) \quad \text{integral spin} \end{aligned}\right\} \tag{15.2}$$

In the quantum field theory this relationship is explained in terms of Pauli's theorem. Thus, to describe the symmetry of the wave functions of a system of identical particles, we use two simple one-dimensional representations, namely the antisymmetric and symmetric, out of the entire set of irreducible representations of the permutation group. We shall usually be concerned with systems of electrons and hence with antisymmetric wave functions. The chief application of the theory of representations of the permutation group in the quantum mechanics of many-particle systems involves an approximation in which the Hamiltonian is assumed to be spin-independent.

The time-independent Schroedinger equation for multi-electron systems is then of the form

$$H(\boldsymbol{r}_1, \boldsymbol{r}_2, \ldots, \boldsymbol{r}_n)\psi(\boldsymbol{r}_1, \ldots, \boldsymbol{r}_n) = E\psi(\boldsymbol{r}_1, \ldots, \boldsymbol{r}_n) \tag{15.3}$$

The complete 'antisymmetric' wave function can be constructed by antisymmetrizing the product of the solution of this equation and the spin functions $\chi(\sigma_1, \sigma_2, \ldots, \sigma_n)$:

$$\Psi(\boldsymbol{r}_1, \sigma_1, \ldots, \boldsymbol{r}_n, \sigma_n) = \sum_p (-1)^{\varepsilon(p)} \psi(\boldsymbol{r}_{p_1}, \boldsymbol{r}_{p_2}, \ldots, \boldsymbol{r}_{p_n}) \chi(\sigma_{p_1}, \ldots, \sigma_{p_n}) \tag{15.4}$$

where $\varepsilon(p)$ is the parity of the permutation p and the sum is evaluated over all the permutations of the n subscripts. The coordinate wave function $\psi(\boldsymbol{r}_1, \ldots, \boldsymbol{r}_n)$, which is the solution of (15.3) is not necessarily antisymmetric with respect to the interchange of its arguments. Since the Hamiltonian $H(\boldsymbol{r}_1, \boldsymbol{r}_2, \ldots, \boldsymbol{r}_n)$ is invariant under any permutation of the variables $\boldsymbol{r}_1, \boldsymbol{r}_2, \ldots, \boldsymbol{r}_n$, the wave functions $\psi(\boldsymbol{r}_1, \boldsymbol{r}_2, \ldots, \boldsymbol{r}_n)$ must, by Wigner's theorem, transform in accordance with one of the irreducible representations of the permutation group S_n. The only limitation which restricts the class of admissible irreducible representations is that the result of the antisymmetrization in (15.4) must be non-zero. The derivation of the wave function for a multi-electron system will be discussed in detail in Chapter 16 but, for the moment, let us investigate the group S_n and all its irreducible representations.

15.2 The permutation group S_n

The elements of the permutation group S_n are the $n!$ permutations of n symbols. A permutation s in which a symbol labelled i is replaced by a symbol labelled s_i is written in the form

$$s=\begin{pmatrix} 1 & 2 & \ldots & n \\ s_1 & s_2 & \ldots & s_n \end{pmatrix}=\begin{pmatrix} i \\ s_i \end{pmatrix} \tag{15.5}$$

The product of two permutations $s=\begin{pmatrix} i \\ s_i \end{pmatrix}$ and $t=\begin{pmatrix} j \\ t_j \end{pmatrix}$ is defined as the permutation

$$ts=\begin{pmatrix} i \\ t_{s_i} \end{pmatrix} \tag{15.6}$$

The identity element of the group is the permutation $\begin{pmatrix} i \\ i \end{pmatrix}$ and the inverse element is defined by

$$s^{-1}=\begin{pmatrix} s_i \\ i \end{pmatrix} \tag{15.7}$$

A cyclic permutation of m symbols ($m \leqslant n$) is one in which a symbol labelled f_k is replaced by a symbol labelled f_{k+1} if $k < m$, while a symbol labelled f_m is replaced by a symbol labelled f_1. Such a permutation will be written in the form of the bracket $(f_1, f_2, \ldots, f_m)$. An arbitrary permutation can be expressed as a product of cyclic permutations. Let us explain this by an example:

$$\begin{pmatrix} 1 & 2 & 3 & 4 & 5 & 6 & 7 & 8 \\ 2 & 4 & 6 & 1 & 5 & 7 & 3 & 8 \end{pmatrix} = (124)\,(367)\,(5)\,(8) \qquad (15.8)$$

The cyclic structure of a permutation will be characterized by a set of numbers $\alpha_1, \alpha_2, \ldots, \alpha_n$, where α_l is the number of cyclic permutations of l symbols contained in the given permutation. For example, the cyclic structure of the identity permutation is characterized by the set

$$\alpha_1 = n, \quad \alpha_2 = \alpha_3 = \ \ldots \ = \alpha_n = 0$$

The cyclic structure of the permutation given by (15.8) is defined by the following values and symbols:

$$\alpha_1 = 2, \quad \alpha_2 = 0, \quad \alpha_3 = 2, \quad \alpha_4 = \alpha_5 = \ \ldots \ = \alpha_8 = 0$$

Since the number of permuted symbols is n, we always have

$$\alpha_1 + 2\alpha_2 + \ \ldots \ + n\alpha_n = n \qquad (15.9)$$

Consider the following two permutations which have the same cyclic structure:

$$a = (123)\,(45)\,(6), \qquad b = (431)\,(26)\,(5)$$

These permutations differ only by the labelling of the symbols, which can be altered with the aid of the permutation

$$t = \begin{pmatrix} 1 & 2 & 3 & 4 & 5 & 6 \\ 4 & 3 & 1 & 2 & 6 & 5 \end{pmatrix}$$

in the sense that $a = t^{-1}bt$. Consequently, a and b belong to the same class. It is clear that, conversely, permutations belonging to a given class, i.e. those that do not differ from one another by the labelling of the symbols, have the same cyclic structure. We thus reach the important conclusion that permutations with the same cyclic structure form a class and, consequently, the number of different classes is equal to the number of different cyclic structures. The number of symbols in a cyclic permutation will be called the length of the cycle. If we denote by $\lambda_1, \lambda_2, \ldots, \lambda_k$ $(\lambda_1+\lambda_2+\ldots+\lambda_k=n)$ the lengths of cycles into which an arbitrary permutation can be decomposed, then the number of different cycle structures, which coincides with the number of classes, will be equal to the number of different possible sub-divisions of n into integral positive parts. From the general theory of representations of finite groups it follows that the permutation group S_n has the same number of different irreducible representations.

15.3 Irreducible representations of the group S_n

Let us now go on to find all the irreducible representations of the group S_n. We have already mentioned two one-dimensional and, hence, irreducible representations, namely the symmetric and antisymmetric, which govern the transformation of the wave functions of particles with integral and half-integral spins, respectively. In the symmetric representation we associate 1 with any permutation, while in the antisymmetric representation we associate $(-1)^{\varepsilon(p)}$ with any permutation, where $\varepsilon(p)$ is the parity of the permutation p, i.e. in this representation we associate 1 with all the even and -1 with all the odd permutations.

We can find the irreducible representations as follows. Consider a function $F(y_1, y_2, \ldots, y_n)$ which is the product of n independent functions, each of which depends on a single argument:

$$F(y_1, y_2, \ldots, y_n)=\psi_1(y_1)\psi_2(y_2)\ldots\psi_n(y_n) \qquad (15.10)$$

Let us now define the permutation operation as applied to the arguments of the function $F(y_1, y_2, \ldots, y_n)$ by the formula

$$\hat{p}F=F(y_{p_1}, y_{p_2}, \ldots, y_{p_n})\equiv F_p(y_1, \ldots, y_n) \qquad (15.11)$$

If we apply all the $n!$ permutations to the function F, we obtain $n!$ independent functions F_p. By applying an arbitrary permutation q to the function F_p we obtain

$$\hat{q}F_p(y_1, y_2, \ldots, y_n) = \hat{q}F(y_{p_1}, y_{p_2}, \ldots, y_{p_n})$$
$$= F\left(y_{q_{p_1}}, y_{q_{p_2}}, \ldots, y_{q_{p_n}}\right) = F_{qp} \tag{15.12}$$

from which it follows that the functions F_p form the basis of a regular representation of the permutation group. We recall that the regular representation is reducible and each irreducible representation of order l is contained in it l times.

Consider a linear $n!$-dimensional space R, whose elements are the arbitrary linear combinations of the functions F_p

$$\sum_{p \in S_n} c_p F_p \tag{15.13}$$

The decomposition of the regular representation which is realized on the functions F_p reduces to the decomposition of the space R into irreducible invariant sub-spaces of S_n. We shall now use the fact that the function F was chosen above in the form of the product (15.10). This will enable us to consider the operator $\hat{P}$ defined by

$$\hat{P}F = \psi_{p_1}(y_1)\psi_{p_2}(y_2) \cdots \psi_{p_n}(y_n) = \hat{p}^{-1}F \tag{15.14}$$

It is readily seen that the operators $\hat{q}$ and $\hat{P}$ commute, i.e.

$$\hat{q}\hat{P} = \hat{P}\hat{q} \tag{15.15}$$

Application of the vector $\hat{P}$ to the function F_q is equivalent to the application of the operator $\hat{p} = \hat{q}\hat{p}^{-1}\hat{q}^{-1}$ to this function. In fact,

$$\hat{P}F_q = \psi_{p_1}(y_{q_1})\psi_{p_2}(y_{q_2}) \cdots \psi_{p_n}(y_{q_n}) \tag{15.16}$$

On the other hand,

$$\hat{q}\hat{p}^{-1}\hat{q}^{-1}F_q = \hat{q}\hat{p}^{-1}\hat{q}^{-1}\psi_1(y_{q_1})\psi_2(y_{q_2}) \cdots \psi_n(y_{q_n})$$
$$= \hat{q}\hat{p}^{-1}\psi_1(y_1)\psi_2(y_2) \cdots \psi_n(y_n) = \hat{q}\psi_{p_1}(y_1)\psi_{p_2}(y_2) \cdots \psi_{p_n}(y_n)$$
$$= \psi_{p_1}(y_{q_1})\psi_{p_2}(y_{q_2}) \cdots \psi_{p_n}(y_{q_n}) \tag{15.17}$$

We see that the application of $\hat{P}$ to the different functions F_q is equivalent to different permutations of the arguments. The result of the application of $\hat{P}$ to an arbitrary element (15.13) of the space R cannot therefore be expressed in terms of any permutation of the arguments.

We shall show that we can use the operators $\hat{P}$ to construct an invariant sub-space of R. Let us select λ functions out of the functions ψ_i. All the permutations of the numbers of these functions form the groups S_λ. Consider the symmetrization operator

$$\hat{\Omega}_{S\lambda} \equiv \sum_{p \in S_\lambda} \hat{P} \tag{15.18}$$

If we apply this operator to all the functions F_p we obtain $n!$ functions which are symmetric with respect to the permutations of the λ functions ψ_i. It is clear that not all the functions constructed in this way will be linearly independent and, consequently, they will form only a sub-space R' of R. We shall show that this sub-space is invariant under any permutation of the arguments $y_1, y_2, \ldots, y_n$. In fact, an arbitrary element of the space R' can be written in the form

$$\hat{\Omega}_{S\lambda} \sum_{q \in S_n} c_q F_q = \sum_{q \in S_n} c_q \sum_{p \in S_\lambda} \hat{P} F_q \tag{15.19}$$

If we apply to it an operator resulting in an arbitrary permutation t of the arguments, and use the commutation rule (15.15), we obtain

$$\hat{t} \sum_{q \in S_n} c_q \sum_{p \in S_\lambda} \hat{P} F_q = \sum_{q \in S_n} c_q \sum_{p \in S_\lambda} \hat{P} F_{tq} = \Omega_{S\lambda} \sum_{q' \in S_n} c'_{q'} F_{q'} \tag{15.20}$$

where $c'_{q'} = c_{t^{-1}q}$.

We see that the function on the right-hand side of this equation belongs to the sub-space R' and, consequently, this sub-space is invariant under any permutation of the arguments. In particular, if $\lambda = n$, the sub-space R' consists of the single function F_S which is symmetric with respect to any permutation of the functions ψ_i:

$$F_S = \hat{\Omega}_{S_n} F = \sum_{p \in S_n} \hat{P} F \qquad (15.21)$$

Similarly, we can derive the invariant sub-space which is antisymmetric with respect to the permutations of the λ functions ψ_i. The corresponding antisymmetrization operator is

$$\hat{\Omega}_{A\lambda} \equiv \sum_{p \in S_\lambda} (-1)^{e(p)} \hat{P} \qquad (15.22)$$

If $\lambda = n$, we obtain the function which is antisymmetric with respect to the group S_n:

$$F_A = \hat{\Omega}_{An} F = \sum_{p \in S_n} (-1)^{e(p)} \hat{P} F \qquad (15.23)$$

To isolate the invariant sub-space, we can carry out successively the symmetrization and antisymmetrization of different sets of functions ψ_i. We cannot, of course, carry out antisymmetrization on functions for which the initial symmetrization was performed, since we would thus obtain zero.

We shall now show that for a certain definite selection of the symmetrization and antisymmetrization operations we can divide the entire space R into invariant sub-spaces which transform by irreducible representations. To indicate the numbers of the functions ψ_i on which symmetrization and antisymmetrization has to be carried out, let us set up tableaux (Young's tableaux) consisting of n cells. The cells are arranged in rows containing $\lambda_1, \lambda_2, \ldots, \lambda_k$ $(\lambda_1 + \lambda_2 + \ldots + \lambda_k = n)$ cells each and arranged one under the other in a non-increasing order as follows:

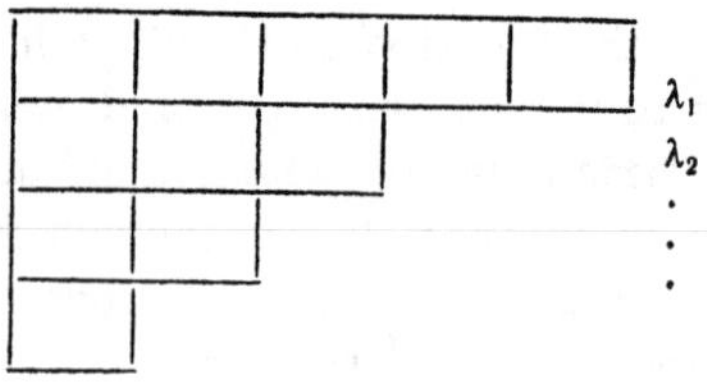

Let us fill the Young's tableau with the numbers of the functions ψ_i. For example, for $n = 13$

1	2	3	4	5
6	7	8	9	
10	11			
12	13			

Let us first carry out antisymmetrization using the functions whose numbers lie in the columns of the tableau and then symmetrization using the numbers in the rows.

There is a device which will enable us to determine the symmetrization and antisymmetrization operations which divide the space R into invariant irreducible subspaces. This device has a recursive character, and reduces to finding the corresponding Young's tableaux with a certain definite occupation of the cells. Let us begin with $n = 1$. Here we can construct only one Young's tableau consisting of a single cell. When $n = 2$, there are two possible Young's tableaux consisting of one row and one column. For $n = 3$ there is a third cell containing the number 3. When $n = 4$, we add a fourth cell, and so on. This method can be illustrated as follows:

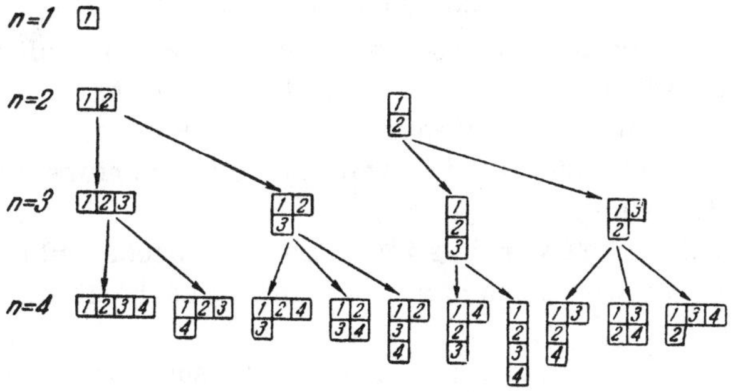

and so on.

With each Young's tableau we associate the Young's operator

$$\hat{\Omega}(\lambda_1, \lambda_2, \ldots, \lambda_k) = \sum_q \hat{Q} \sum_p (-1)^{\varepsilon(p)} \hat{P}$$

where q runs over all the permutations of the numbers in each row, and p all the permutations in each column. In view of the foregoing discussion, we conclude that each such Young's operator isolates a certain invariant sub-space in the space R. We shall show that these invariant sub-spaces are independent. We note that by counting their number and comparing the result with the known number of irreducible representations into which the regular representation decomposes, we necessarily conclude that in each of the sub-spaces which we have constructed only one of the irreducible representations of S_n is realized.

We shall therefore show that invariant sub-spaces constructed with the aid of Young's operators, for which the tableaux are found by the above recurrent method, have no common elements. Our proof will be heuristic and will be based on two facts which the reader will readily verify by considering Young's tableaux:

1. Repeated application of the same Young's operator to any element of the space R yields a non-zero result.

2. Successive application of two different Young's operators to any element of the space R yields zero. This is readily verified, at least for a given order of application of the operators. In fact, it is readily shown that one of this pair of operators will always contain symmetrization with respect to the functions for which the other performs anti-symmetrization.

If the corresponding invariant sub-spaces had common elements we would reach a contradiction of the above statements.

Finally, it is readily seen that the sub-spaces corresponding to Young's tableaux with the same structure $\{\lambda\} = \lambda_1, \lambda_2, \ldots, \lambda_k$ but differing only by the occupation of the cells transform in accordance with equivalent representations. Hence, it follows that the number of inequivalent representations is equal to the number of Young's tableaux with different structure, or the number of sub-divisions of n into integral positive terms $\lambda_1 + \lambda_2 + \ldots + \lambda_k = n$, i.e. the number of classes. Inspection of our table will readily show that for the number $r_{\{\lambda\}}$ of equivalent representations we must always have

$$\sum_{\{\lambda\}} r^2_{\{\lambda\}} = n! \tag{15.24}$$

Comparison of this with (3.81) leads to the conclusion that the numbers $r_{\{\lambda\}}$ are the orders of the irreducible representations $l_{\{\lambda\}}$: $r_{\{\lambda\}} = l_{\{\lambda\}}$. Using the theorem on the decomposition of a regular representation into irreducible parts we may conclude that the number of irreducible representations which it contains is equal to $\sum_{\{\lambda\}} l_{\{\lambda\}}$, but since $\sum_{\{\lambda\}} l_{\{\lambda\}} = \sum_{\{\lambda\}} r_{\{\lambda\}}$, this number is equal to the number of invariant sub-spaces which we have constructed. Hence, it follows that one of the irreducible representations of the permutation group S_n is realized in each of these sub-spaces. We shall denote these irreducible representations by the symbols $\Delta_{\{\lambda\}}$.

Consider now the symmetry properties of the functions belonging to the basis of the irreducible representation $\Delta_{\{\lambda\}}$ To be specific, consider the function $F_{\{\lambda\}}$ which is obtained as a result of the application of the operator $\hat{\Omega}_{\{\lambda\}}$ to the function $F = \psi_1(y_1) \ldots \psi_n(y_n)$:

$$F_{\{\lambda\}} = \hat{\Omega}_{\{\lambda\}} F = \sum_q \hat{Q} \sum_p (-1)^{\varepsilon(p)} \hat{P} F \tag{15.25}$$

From this definition it follows that the function $F_{\{\lambda\}}$ should be symmetric under the permutations of the functions ψ_i whose numbers lie in the rows of Young's tableau (i.e. with respect to the operators $\hat{Q}$). However, we cannot say that the function $F_{\{\lambda\}}$ will be antisymmetric with respect to the operators $\hat{P}$, since $\hat{P}$ and $\hat{Q}$ do not commute. However, according to (15.14), we can write

$$\hat{P}F = \hat{p}^{-1}F \tag{15.26}$$

and, therefore, the function $F_{\{\lambda\}}$ can also be written in the form

$$F_{\{\lambda\}} = \sum_q \hat{Q} \sum_p (-1)^{\varepsilon(p)} \hat{p}F \tag{15.27}$$

Since the operators $\hat{Q}$ and $\hat{p}$ commute, we have

$$F_{\{\lambda\}} = \sum_p (-1)^{\varepsilon(p)} \hat{p} \sum_q \hat{Q}F \tag{15.28}$$

and, finally, using (15.26) once again we have

$$F_{\{\lambda\}} = \sum_p (-1)^{\varepsilon(p)} \hat{p} \sum_q \hat{q}F \tag{15.29}$$

Thus, instead of performing antisymmetrization and then symmetrization with respect to the numbers of the functions ψ_i, we can first symmetrize and then antisymmetrize with respect to the numbers of the corresponding arguments. This shows, in particular, that our initial assumption that the function F can be represented by a product is not a basic one but is of an auxiliary character only. From (15.28) or (15.29) it follows that the function $F_{\{\lambda\}}$ should be antisymmetric with respect to the interchange of its arguments whose numbers lie in the columns of Young's tableau.

We can now summarize our results as follows. An irreducible representation $\Delta_{\{\lambda\}}$ of the permutation group is determined by the subdivision of n into integral positive parts:

$$n = \lambda_1 + \lambda_2 + \ldots + \lambda_k, \quad \lambda_1 \geqslant \lambda_2 \geqslant \ldots \geqslant \lambda_k$$

The regular representation of the group S_n is realized in the linear space R formed by the functions $F(y_{p_1}, \ldots, y_{p_n})$, where $(p_1\ p_2 \ldots p_n)$ are all the possible permutations of the numbers $1, 2, \ldots, n$. Without restricting the generality of our results, we can represent the function $F(y_1, y_2, \ldots, y_n)$ as the product of the n functions $\psi_1(y_1) \ldots \psi_n(y_n)$. The decomposition of this representation into irreducible parts can be carried out with the aid of Young's operators corresponding to Young's tableaux constructed by the recursion rule.

Exercises

15. 1. Show that the characters of the representations of the permutation group must be real.

15. 2. Determine the number of elements in a class with the cyclic structure $(\alpha_1\ \alpha_2 \ldots \alpha_m)$.

15. 3. Show that the parity of a permutation is determined by the parity of the sum $\alpha_2 + \alpha_4 + \alpha_6 + \ldots$

Chapter 16

Symmetrized Powers of Representations

In this chapter we shall introduce the concept of a symmetrized power of a representation. This will form the basis of most of the applications of the permutation group to problems in quantum mechanics.

16.1 Vectors and tensors in *n*-dimensional space

Consider an n-dimensional vector space R_n. The components of a vector in this space in a certain chosen basis will be denoted by V_i, $i = 1, 2, \ldots, n$. Suppose that a representation of a group G is given in the space R_n. The matrices of the representations corresponding to the elements $a, b, \ldots$ of G will be denoted by $\| a_{jj'} \|$, $\| b_{jj'} \|$, ... Therefore, we can associate with an element a the following transformation of vectors in R_n:

$$V'_j = \sum_{j'=1}^{n} a_{jj'} V_{j'} \tag{16.1}$$

The representation matrix of the unity element will be denoted by $\| e_{jj'} \|$, $e_{jj'} = \delta_{jj'}$.

In addition to the vectors, let us consider the tensor $V^{(m)}$ of rank m in the space R_n, the components of which will be denoted by $V_{j_1 j_2 \dots j_m}$, $j_k = 1, 2, \dots, n$. To each transformation (16.1) there corresponds the following transformation of the tensor:

$$V'_{j_1 j_2 \dots j_m} = \sum_{j'_1, \dots, j'_m} a_{j_1 j'_1} \dots a_{j_m j'_m} V_{j'_1 j'_2 \dots j'_m} \tag{16.2}$$

The matrix of this transformation is the m-fold direct product of the matrix $\|a_{jj'}\|$. We shall denote it by $(a)^m$ or $\left\| a_{j_1 j_2 \dots j_m, j'_1 j'_2 \dots j'_m} \right\|$, where

$$a_{j_1 j_2 \dots j_m, j'_1 j'_2 \dots j'_m} = a_{j_1 j'_1} a_{j_2 j'_2} \dots a_{j_m j'_m} \tag{16.3}$$

The matrices $(a)^m$ form a representation of the group G.

It will be convenient to introduce the matrix $p(a)^m$ with the elements

$$\{p(a)^m\}_{j_1 j_2 \dots j_m, j'_1 j'_2 \dots j'_m} = a_{j_{p_1} j_{p_2} \dots j_{p_m}, j'_1 j'_2 \dots j'_m} \tag{16.4}$$

where $(p_1 p_2 \dots p_m)$ is a permutation of the numbers $1, 2, \dots, m$. We note that the matrices $(a)^m$ have the following property:

$$a_{j_{p_1} j_{p_2} \dots j_{p_m}, j'_{p_1} j'_{p_2} \dots j'_{p_m}} = a_{j_1 j_2 \dots j_m, j'_1 j'_2 \dots j'_m} \tag{16.5}$$

In fact, an identical permutation of the first and second indices of the elements of the matrix $(a)^m$ reduces the permutation of the factors in (16.3).

16.2 Permutation matrices for tensor indices

Consider the tensor $pV^{(m)}$ whose components are defined by

$$pV_{j_1 \dots j_m} = V_{j_{p_1} \dots j_{p_m}} \tag{16.6}$$

We shall show that the components of this tensor can be expressed in terms of the components of the tensor $V_{j_1 j_2 \dots j_m}$ with the aid of the matrix $p(e^m)$, where $\|e_{ik}\|$ is the unit matrix of order n and the elements $(e)^m$ are, obviously, of the form

$$e_{j_1 j_2 \ldots j_m,\; j_1' j_2' \ldots j_m'} = \delta_{j_1 j_1'} \delta_{j_2 j_2'} \ldots \delta_{j_m j_m'} \tag{16.7}$$

In fact, we can write

$$\sum_{j_1', \ldots, j_m'} \{p\,(e)^m\}_{j_1 \ldots j_m,\; j_1' \ldots j_m'} V_{j_1' \ldots j_m'}$$
$$= \sum_{j_1', \ldots, j_m'} \delta_{j_{p_1} j_1'} \ldots \delta_{j_{p_m} j_m'} V_{j_1' \ldots j_m'} = V_{j_{p_1} \ldots j_{p_m}} \tag{16.8}$$

We shall show that the matrices $p\,(e)^m$ form a representation of the permutation group S_n, i.e.

$$q\,(e)^m\, p\,(e)^m = qp\,(e)^m \tag{16.9}$$

where q and p are arbitrary permutations of the numbers $1, 2, \ldots, m$. In fact, there is the more general equation

$$q\,(b)^m\, p\,(a)^m = qp\,(ba)^m \tag{16.10}$$

which we shall need later and prove below. It is clear that (16.9) is a special case of (16.10) with $a = b = e$. To prove (16.10), we shall write out the $(j_1 j_2 \ldots j_m,\; k_1 k_2 \ldots k_m)$-th element of the matrix on the left of (16.10). We obtain

$$\sum_{j'} q\,(b)_{j_1 \ldots j_m,\; j_1' \ldots j_m'}\, p\,(a)_{j_1' \ldots j_m',\; k_1 \ldots k_m}$$
$$= \sum_{j'} b_{j_{q_1} \ldots j_{q_m},\; j_1' \ldots j_m'}\, a_{j_{p_1}' \ldots j_{p_m}',\; k_1 \ldots k_m} \tag{16.11}$$

We shall now use the fact that the same permutation of the first and second indices in accordance with (16.5) will leave the matrix $(b)^m$ unaltered. We can therefore write

$$b_{j_{q_1} \ldots j_{q_m},\; j_1' \ldots j_m'} = b_{j_{q_{p_1}} \ldots j_{q_{p_m}},\; j_{p_1}' \ldots j_{p_m}'} \tag{16.12}$$

Equation (16.11) can then be continued as follows:

$$\sum_{j} b_{j_{q_{p_1}} \ldots j_{q_{p_m}},\; j_{p_1}' \ldots j_{p_m}'}\, a_{j_{p_1}' \ldots j_{p_m}',\; k_1 \ldots k_m}$$
$$= (ba)_{j_{q_{p_1}} \ldots j_{q_{p_m}},\; k_1 \ldots k_m} = \{qp\,(ba)\}_{j_1 j_2 \ldots j_m,\; k_1 \ldots k_m} \tag{16.13}$$

This is, in fact, the $(j_1 \ldots j_m, k_1 \ldots k_m)$-th element of the matrix on the right-hand side of (16.10), and this proves (16.10) and hence (16.9).

16.3 Relationship between the representations of the group S_n and the group G in tensor space

We have defined the representation of the permutation group by the matrices $p(e)^m$ and the representation of the group G by the matrices $(a)^m$. The basis of these representations is the basis of the tensor space, i.e. the set of n^m tensors for which only one component is non-zero and equal to unity. We shall denote the basis vectors by

$$\eta_{i_1 i_2 \ldots i_m}$$

An arbitrary tensor $V^{(m)}$ with the components $V_{j_1 \ldots j_m}$ can then be written in the form

$$V^{(m)} = \sum_{i_1, \ldots, i_m} V_{i_1 \ldots i_m} \eta_{i_1 \ldots i_m} \tag{16.14}$$

We shall now show that the matrices $p(e)^m$ and $(a)^m$ commute. In fact, using (16.10), we have

$$q(e)^m (a)^m = q(a)^m, \quad (a)^m q(e)^m = q(a)^m \tag{16.15}$$

and hence

$$q(e)^m (a)^m = (a)^m q(e)^m \tag{16.16}$$

Let us now transform in the tensor space to a new basis, choosing it so that a representation of the permutation group decomposes into irreducible parts. The matrices of the reduced representation will be written in the form

$$\widetilde{q(e)}^m = \sum_{\{\lambda\}}^{\oplus} \{E_{r_{\{\lambda\}}} \times \Delta_{\{\lambda\}}(q)\} \tag{16.17}$$

where $r_{\{\lambda\}}$ is the multiplicity of the irreducible representation $\Delta_{\{\lambda\}}$ in the representation $q(e)^m$, and $E_{r_{\{\lambda\}}}$ is the unit matrix of order $r_{\{\lambda\}}$. The matrices $\widetilde{q(e)}^m$ determine the new transformation of the components of the tensor written in the new basis. The new basis vectors will be denoted by $\eta_{\{\lambda\}\varkappa\alpha}$, where the symbol $\{\lambda\}$ labels the irreducible representations of the permutation group, $\varkappa$ distinguishes repeating equivalent irreducible representations $\varkappa = 1, 2, \ldots, r_{\{\lambda\}}$ and α identifies the basis vectors of the irreducible representations $\alpha = 1, 2, \ldots, l_{\{\lambda\}}$. The new basis vectors can be expressed in terms of the old as follows:

$$\eta_{\{\lambda\}\varkappa\alpha} = \sum_{j_1, j_2, \ldots, j_n} T^{(\{\lambda\}\alpha\varkappa)}_{j_1 j_2 \cdots j_n} \eta_{j_1 j_2 \cdots j_n} \tag{16.18}$$

The coefficients $T^{(\{\lambda\}\alpha\varkappa)}_{j_1 j_2 \cdots j_n}$ can be regarded as the components of the tensor $\eta_{\{\lambda\}\alpha\varkappa}$ in the old basis.

Let us now find the form of the matrices $(a)^m$ in the new basis. Using the commutation property (16.16) and Schur's lemma, we obtain (see Chapter 5)

$$(\tilde{a})^m = \sum_{\{\lambda\}}^{\oplus} \left\{ R^{\{\lambda\}}(a) \times E_{l_{\{\lambda\}}} \right\} \tag{16.19}$$

or

$$(\tilde{a})^m_{\{\lambda\}\varkappa\alpha,\, \{\lambda'\}\varkappa'\alpha'} = \delta_{\{\lambda\}\{\lambda'\}} \delta_{\alpha\alpha'} R^{\{\lambda\}}_{\varkappa\varkappa'}(a) \tag{16.20}$$

The order of the matrices $R^{\{\lambda\}}(a)$ is $r_{\{\lambda\}}$. They form a representation of the group a in the space with basis vectors $\eta_{\{\lambda\}\varkappa\alpha}$, $\varkappa = 1, 2, \ldots, r_{\{\lambda\}}$ (for fixed $\{\lambda\}$ and α). The representation $R^{\{\lambda\}}(a)$ is called the symmetrized m-th power of the representation a corresponding to the irreducible representation $\Delta_{\{\lambda\}}$ of the permutation group.

16.4 Characters of a symmetrized power of representations

In applications one often encounters the problem of the decomposition of the representation $R_{\{\lambda\}}(a)$ into irreducible representations. This decomposition can be obtained if we know the characters of the representation.

Let us first solve an auxiliary problem. Consider the trace of the matrix $q(a)^m$ and suppose that the permutation q has the cyclic structure $(\alpha_1\alpha_2\ldots\alpha_m)$. If the permutation consisted of a single cycle of length m, we would find that

$$\{q(a)\}_{j_1\ldots j_m,\, j_1'\ldots j_m'}$$
$$= a_{j_m j_1 j_2\ldots j_{m-1},\, j_1'\ldots j_m'} = a_{j_m j_1'} a_{j_1 j_2'}\ldots a_{j_{m-1} j_m'} \quad (16.21)$$

Consequently,

$$\operatorname{Sp} q(a) = \sum_{j_1,\ldots,j_m} a_{j_m j_1} a_{j_1 j_2}\ldots a_{j_{m-1} j_m} = \operatorname{Sp}(a)^m \quad (16.22)$$

and hence it is readily shown that for an arbitrary permutation we have

$$\operatorname{Sp} q(a)^m = (\operatorname{Sp}(a))^{\alpha_1}(\operatorname{Sp}(a)^2)^{\alpha_2}\ldots(\operatorname{Sp}(a)^m)^{\alpha_m} \quad (16.23)$$

Let us now establish the relation between the traces of the matrices $q(a)$ and $R^{\{\lambda\}}(a)$. To do this, we shall use the fact that, according to (16.15),

$$q(a)^m = q(e)^m (a)^m$$

or

$$\widetilde{q(a)}^m = \widetilde{q(e)}^m \widetilde{(a)}^m \quad (16.24)$$

and, consequently, we find from (16.17) and (16.19) that

$$\widetilde{q(a)}^m = \sum_{\{\lambda\}}^{\oplus} \{E_{r_{\{\lambda\}}} \times \Delta_{\{\lambda\}}(q)\}\{R^{\{\lambda\}}(a) \times E_{l_{\{\lambda\}}}\}$$
$$= \sum_{\{\lambda\}}^{\oplus} \{R^{\{\lambda\}}(a) \times \Delta_{\{\lambda\}}(q)\} \quad (16.25)$$

Hence, we find that

$$\operatorname{Sp} q(a)^m = \operatorname{Sp} \widetilde{q(a)}^m = \sum_{\{\lambda\}} \operatorname{Sp} R^{\{\lambda\}}(a) \operatorname{Sp} \Delta_{\{\lambda\}}(q) \quad (16.26)$$

Using the orthogonality of the characters $\chi^{\{\lambda\}}(q)$ of irreducible representations of the permutation group, we find that

$$\operatorname{Sp} R_{\{\lambda\}}(q) = \frac{1}{m!}\sum_q \operatorname{Sp} q(a)^m \chi^{\{\lambda\}}(q) \quad (16.27)$$

We note that characters of the representations of the permutation group are real. Substituting (16.23) into (16.20) and transforming from summation over the group elements to summation over classes, we obtain

$$\operatorname{Sp} R^{\{\lambda\}}(a)=\frac{1}{m!}\sum_{\{\alpha\}} k_{\{\alpha\}}\chi^{\{\lambda\}}_{\{\alpha\}}(\operatorname{Sp} a)^{\alpha_1}\dots(\operatorname{Sp} a^m)^{\alpha_m} \tag{16.28}$$

where $\chi^{\{\lambda\}}_{\{\alpha\}}$ denotes the character of the irreducible representation $\Delta_{\{\lambda\}}(q)$ corresponding to the class $\{\alpha\}$, and $k_{\{\alpha\}}$ is the number of elements in the class. Finally,

$$k_{\{\alpha\}}=\frac{m!}{\alpha_1!\,2^{\alpha_2}\alpha_2!\dots m^{\alpha_m}\alpha_m!} \tag{16.29}$$

One very frequently has to deal with symmetric or antisymmetric powers of representations, i.e. with representations $R_m(a)$ and $R_{\{1\,1\dots1\}}(a)$ which correspond to identity and antisymmetric irreducible representations of the permutation group. The characters of the identity representation are equal to unity, $\chi^{(m)}_{\{\alpha\}}=1$. Consequently, for the characters of the symmetric power we obtain

$$\operatorname{Sp} R_m(a)=\sum_{\{\alpha\}}\frac{(\operatorname{Sp} a)^{\alpha_1}(\operatorname{Sp} a^2)^{\alpha_2}\dots(\operatorname{Sp} a^m)^{\alpha_m}}{\alpha_1!\,2^{\alpha_2}\alpha_2!\dots m^{\alpha_m}\alpha_m!} \tag{16.30}$$

The characters of the antisymmetric representation can be written in the form

$$\chi^{\{1\,1\dots1\}}_{\{\alpha\}}=(-1)^{\alpha_2+\alpha_4+\dots} \tag{16.31}$$

Substituting this into (16.28), we obtain the following expression for the characters of the antisymmetric power of a representation:

$$\operatorname{Sp} R_{\{1\,1\dots1\}}(a)=\sum_{\{\alpha\}}(-1)^{\alpha_2+\alpha_4+\dots}\frac{(\operatorname{Sp} a)^{\alpha_1}(\operatorname{Sp} a^2)^{\alpha_2}\dots(\operatorname{Sp} a^m)^{\alpha_m}}{\alpha_1!\,2^{\alpha_2}\alpha_2!\dots m^{\alpha_m}\alpha_m!} \tag{16.32}$$

Exercises

16. 1. Write down expressions for the characters of symmetric and antisymmetric squares of a representation.

16. 2. Find the decomposition into irreducible representations of the symmetric second and third powers of irreducible representations of the group O. Compare the results with the decomposition of direct products of the corresponding representations.

16. 3. Find representations governing the transformation of the vibrational wave function of the octahedral molecule considered in Chapter 6 for all the two-quantum excited states.

Chapter 17

Symmetry Properties of Multi-Electron Wave Functions

In Section 15.1 we outlined a plan for a group-theoretical investigation of a multi-electron system. We shall now try to execute this plan, using the theory of the permutation group developed in Chapters 15 and 16.

17.1 Formulation of the problem

The eigenfunctions of the Schroedinger equation

$$H(\boldsymbol{r}_1, \ldots, \boldsymbol{r}_n)\Psi(\boldsymbol{r}_1, \ldots, \boldsymbol{r}_n) = E\Psi(\boldsymbol{r}_1, \ldots, \boldsymbol{r}_n), \tag{17.1}$$

belonging to a given eigenvalue must transform in accordance with an irreducible representation $\Delta_{\{\lambda\}}$ of the permutation group of the variables $\boldsymbol{r}_1, \boldsymbol{r}_2, \ldots, \boldsymbol{r}_n$. However, the degree of degeneracy of the eigenvalue, which is equal to the order of the irreducible representation, is not equal to the degree of degeneracy (statistical weight) of the corresponding energy level. The degree of degeneracy of an energy level must be determined as the number of complete (i.e. depending on the spins also) antisymmetric functions which are the eigenfunctions of (17.1) belonging to the particular eigenvalue.

Let $\Psi(\boldsymbol{r}_1, \boldsymbol{r}_2, \ldots, \boldsymbol{r}_n)$ be one of the degenerate coordinate eigenfunctions of (17.1). It is clear that

$$\Psi(\boldsymbol{r}_1, \boldsymbol{r}_2, \ldots, \boldsymbol{r}_n)\chi(\sigma_1, \sigma_2, \ldots, \sigma_n) \tag{17.2}$$

where $\chi(\sigma_1, \sigma_2, \ldots, \sigma_n)$ is the spin function (spinor of rank n), is also an eigenfunction of (17.1). If we transpose the arguments in (17.2) so that

$$\Psi(\boldsymbol{r}_{p_1}, \boldsymbol{r}_{p_2}, \ldots, \boldsymbol{r}_{p_n})\chi(\sigma_{p_1}, \sigma_{p_2}, \ldots, \sigma_{p_n}) \tag{17.3}$$

we again obtain an eigenfunction of the Schroedinger equation with the same eigenvalue. This enables us to construct the complete antisymmetric wave function which is the eigenfunction of the Hamiltonian in the form

$$\Psi_A = \sum_p (-1)^{\varepsilon(p)}\Psi(\boldsymbol{r}_{p_1}, \ldots, \boldsymbol{r}_{p_n})\chi(\sigma_{p_1}, \ldots, \sigma_{p_n}) \tag{17.4}$$

We shall see below, however, that an antisymmetric wave function of the form (17.4), which is not identically zero, cannot be constructed for all solutions $\Psi(\boldsymbol{r}_1, \ldots, \boldsymbol{r}_n)$ of (17.1). There is a definite restriction on the irreducible representations $\Delta_{\{\lambda\}}$ which govern the transformation of the solutions of (17.1). To establish these limitations we must first determine the representations $\Delta_{\{\lambda'\}}$ of the permutation group which govern the transformation of the spin functions $\chi(\sigma_1, \sigma_2, \ldots, \sigma_n)$. We shall use the following criterion: the possible irreducible representations $\Delta_{\{\lambda\}}$ are those for which the direct product $\Delta_{\{\lambda\}} \times \Delta_{\{\lambda'\}}$ contains an antisymmetric irreducible representation.

17.2 Symmetry properties of the spin wave functions

The spin function $\chi(\sigma_1, \sigma_2, \ldots, \sigma_n)$ of an n-electron system is a spinor of rank n with respect to rotations of three-dimensional space. We recall that when the coordinate system is rotated through the Euler angles $\varphi_1, \theta, \varphi_2$, the components of a spinor of rank n transform so that

$$\chi'(\sigma_1, \sigma_2, \ldots, \sigma_n) = \sum_{\sigma'} \alpha_{\sigma_1\sigma_1'}[\varphi_1, \theta, \varphi_2] \ldots \alpha_{\sigma_n\sigma_n'}[\varphi_1, \theta, \varphi_2] \chi(\sigma_1', \sigma_2', \ldots, \sigma_n') \qquad (17.5)$$

where the matrix $\|\alpha_{\sigma\sigma'}[\varphi_1, \theta, \varphi_2]\|$ is the canonical matrix of the irreducible representation $D^{(1/2)}$ of the rotation group. The transformation matrices for (17.5) form a representation of the rotation group which is the n-th power of the representation by the matrices α. It is clear that the spinor $\chi(\sigma_1, \sigma_2, \ldots, \sigma_n)$ can be regarded as a tensor of rank n in two-dimensional space, and all the results obtained in the last chapter are valid for it.

In a given set of coordinates we can regard the set of 2^n components of a spinor as the values of the function of n spin variables σ_i, each of which can assume only the two values $\frac{1}{2}$ and $-\frac{1}{2}$. As a result of a permutation p of the arguments of this function, or of the indices of the spinor, we obtain a new function

$$\hat{p}\chi(\sigma_1, \sigma_2, \ldots, \sigma_n) = \chi(\sigma_{p_1}, \sigma_{p_2}, \ldots, \sigma_{p_n}) \equiv \chi_p(\sigma_1, \sigma_2, \ldots, \sigma_n) \qquad (17.6)$$

The quantities $\chi_p(\sigma_1, \sigma_2, \ldots, \sigma_n)$ transform under rotations of three-dimensional space by the rule indicated by (17.5) and are therefore the components of a new spinor. In fact, we have

$$\chi_p'(\sigma_1, \sigma_2, \ldots, \sigma_n) \equiv \chi'(\sigma_{p_1}, \ldots, \sigma_{p_n}) = \sum_{\sigma'} \alpha_{\sigma_{p_1}\sigma_1'} \alpha_{\sigma_{p_2}\sigma_2'} \ldots \alpha_{\sigma_{p_n}\sigma_n'} \chi(\sigma_1', \ldots, \sigma_n') \qquad (17.7)$$

This can also be rewritten in the equivalent form

$$\chi_p'(\sigma_1, \sigma_2, \ldots, \sigma_n) = \sum_{\sigma'} \alpha_{\sigma_{p_1}\sigma_{p_1}'} \alpha_{\sigma_{p_2}\sigma_{p_2}'} \ldots \alpha_{\sigma_{p_n}\sigma_{p_n}'} \chi(\sigma_{p_1}', \ldots, \sigma_{p_n}') \qquad (17.8)$$

and, if we rearrange the factors, we finally obtain

$$\chi'_p(\sigma_1, \ldots, \sigma_n) = \sum_{\sigma'} a_{\sigma_1\sigma'_1} \cdots a_{\sigma_n\sigma'_n} \chi_p(\sigma'_1, \ldots, \sigma'_n) \tag{17.9}$$

It is clear that any linear combination of the form

$$\sum_{p \in S_n} c_p \chi_p (\sigma_1, \sigma_2, \ldots, \sigma_n) \tag{17.10}$$

will also define a spinor.

If we apply to the spinor $\chi(\sigma_1, \ldots, \sigma_n)$ all the permutation operators p belonging to the group S_n we obtain $n!$ spinors. However, not all of these will be independent. The number of independent spinors cannot exceed the number of components of the spinor, which is equal to 2^n. As the independent spinors we can take the set of 2^n spinors, each of which has only one component which is non-zero and equal to unity. The representation $p(e)^n$ of the permutation group, where (e) is the second-order unit matrix, will be realized in the set of spinors thus obtained. The decomposition of the representation $p(e)^n$ into irreducible parts can be carried out by a method analogous to that used in Chapter 15 to decompose a regular representation.

Let us set up a Young's tableau, using the recursion rule established in the last chapter. With each Young's tableau we shall associate the operator

$$\Omega_{\{\lambda\}} = \sum_p (-1)^{\varepsilon(p)} \hat{p} \sum_q \hat{q} \tag{17.11}$$

where summation over p represents summation over all the permutations of the arguments whose numbers lie in the columns of the Young's tableau, while summation over q means summation over all the permutations of the arguments whose numbers lie in the rows of the tableau. Applying the operator (17.11) to the spinor $\chi(\sigma_1, \sigma_2, \ldots, \sigma_n)$, we obtain the spinor $\chi_{\{\lambda\}}$, which belongs to the basis of the irreducible representation $\Delta_{\{\lambda\}}$ of the permutation group of its indices. We note that if Young's tableau consists of more than two rows, then the spinor $\chi_{\{\lambda\}}(\sigma_1, \sigma_2, \ldots, \sigma_n)$ must be antisymmetric with respect to the permutation of at least three indices. However, all the components of this spinor must be zero since, of the three indices, two must necessarily coincide. Young's tableaux consisting of more than two rows can therefore be eliminated from our analysis. Admissible Young's tableaux are of the form

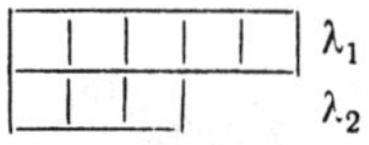

The spinors $\chi_{\{\lambda\}}$, which correspond to Young's tableaux with the same structure, form the basis of an irreducible representation D of the permutation group of its indices.

Let us determine the number $r_{\{\lambda\}}$ of the independent components of the spinor $\chi_{\{\lambda\}}$. Since this spinor must be antisymmetric with respect to the permutation of the indices whose numbers lie in the columns of the corresponding Young's tableau, the only non-zero components will be those for which these indices assume different values. Moreover, all the components which differ only by the value of these indices will be equal (apart from the sign). Since the spinor $\chi_{\{\lambda\}}$ is symmetric with respect to the remaining indices, the number of independent components will be equal to the number of different sets out of the $\lambda_1-\lambda_2$ numbers equal to $\frac{1}{2}$ or $-\frac{1}{2}$. These sets are of the form

$$\left.\begin{array}{cccc} \frac{1}{2} & \frac{1}{2} & \cdots & \frac{1}{2} \\ -\frac{1}{2} & \frac{1}{2} & \cdots & \frac{1}{2} \\ \cdot & \cdot & \cdots & \cdot \\ -\frac{1}{2} & -\frac{1}{2} & \cdots & -\frac{1}{2} \end{array}\right\} \qquad (17.12)$$

Their number is $\lambda_1-\lambda_2+1$. The number of independent components of the spinor $\chi_{\{\lambda\}}$ is therefore $r_{\{\lambda\}}=\lambda_1-\lambda_2+1$. This means that there is an equal number of independent spinors corresponding to a given occupation of the cells in the Young's tableau. It can readily be shown that for each value n we thus obtain 2^n independent spinors.

The independent components of a symmetric spinor $\chi_{\{\lambda\}}$ must transform in accordance with the representation $R_{\{\lambda\}}(\alpha)$ of the rotation group, which is the symmetrized power of the representation α. Let us find the decomposition of the representation $R_{\{\lambda\}}(\alpha)$ into irreducible representations. We know that the basis of an irreducible representation $D^{(l)}$ of the rotation group can be constructed from the $2l+1$ eigenvectors of the infinitesimal operator $\hat{H}_3$ with eigenvalues $-l, -l+1, \ldots, l$. Therefore, to decompose the representation $R_{\{\lambda\}}(\alpha)$ into irreducible representations it is sufficient to establish the multiplicity of the eigenvalues of the operator $\hat{H}_3$ in the basis space of the representation R. In accordance with the transformation rule (17.5), the spinor for which only one component is non-zero (in the given set of coordinates) will be the eigenvector of the infinitesimal operator $\hat{H}_3$, which, in this case, coincides with the operator $\hat{S}_3$ (apart from a factor):

$$\hat{S}_3\chi(\sigma_1, \ldots, \sigma_n) = \hbar\hat{H}_3\chi(\sigma_1, \ldots, \sigma_n)$$

$$= \hbar(\sigma_1+\sigma_2+\ldots+\sigma_n)\chi(\sigma_1, \ldots, \sigma_n) \quad (17.13)$$

Since the antisymmetric indices assume values with opposite signs, while the symmetric indices are determined by (17.12), we obtain the following eigenvalues for the independent components of our spinor:

$$\frac{1}{2}(\lambda_1-\lambda_2),\ \frac{1}{2}(\lambda_1-\lambda_2)-1,\ \ldots,\ -\frac{1}{2}(\lambda_1-\lambda_2) \quad (17.14)$$

Hence, it follows that the components of the spinor $\chi_{\{\lambda\}}$ transform in accordance with the irreducible representation $D^{\left(\frac{\lambda_1-\lambda_2}{2}\right)}$ of the rotation group and, consequently, they are the eigenvectors of the operator $\hat{H}_1^2+\hat{H}_2^2+\hat{H}_3^2$:

$$(\hat{H}_1^2+\hat{H}_2^2+\hat{H}_3^2)\chi_{\{\lambda\}}(\sigma_1, \sigma_2, \ldots, \sigma_n)$$

$$= \frac{\lambda_1-\lambda_2}{2}\left(\frac{\lambda_1-\lambda_2}{2}+1\right)\chi_{\{\lambda\}}(\sigma_1, \sigma_2, \ldots, \sigma_n) \quad (17.15)$$

The most important results for the ensuing analysis are the following. An arbitrary spin function $\chi(\sigma_1, \sigma_2, \ldots, \sigma_n)$ can be expanded in terms of the spin functions $\chi_{\{\lambda\}}(\sigma_1, \sigma_2, \ldots, \sigma_n)$ which transform according to the irreducible representations of the permutation group for its arguments. The admissible irreducible representations will be only those for which the Young's tableau contains not more than two rows $\{\lambda\} = \{\lambda_1, \lambda_2\}$. The components of the spin function $\chi_{\{\lambda\}}$ form the basis of an irreducible representation of the rotation group with the weight $\frac{\lambda_1 - \lambda_2}{2}$ and, consequently, they are the eigenfunctions of the square of the total spin angular momentum operator with the eigenvalue

$$\hbar^2\left(\frac{\lambda_1-\lambda_2}{2}\right)\left(\frac{\lambda_1-\lambda_2}{2}+1\right) \tag{17.16}$$

17.3 Relation between the symmetry of the spin and coordinate wave functions

We have found the representation $p(e)^n$ of the group S_n which governs the transformation of the spin function. This representation can be decomposed into irreducible representations for which Young's tableaux consist of not more than two lines. The irreducible representation $\Delta_{\{\lambda\}}$ which governs the transformation of the coordinate wave function must be such that the direct product $\Delta_{\{\lambda\}} \times p(e)^n$ contains the antisymmetric representation. To determine the admissible representations $\Delta_{\{\lambda\}}$, we shall use the following theorem.

The direct product of two irreducible representations contains an antisymmetric representation only if the corresponding Young's tableaux are transposes of each other, i.e. each is obtained from the other by replacing rows by columns.

The proof of this theorem can be outlined as follows. Consider the direct product $\Delta_{\{\lambda\}} \times \Delta_A$, where Δ_A is an antisymmetric irreducible representation, $\Delta_A \equiv \Delta_{\{1,1,\ldots,1\}}$, and $\Delta_{\{\lambda\}}$ is an arbitrary irreducible representation. Since the representation Δ_A is one-dimensional, it follows that the representation Δ is irreducible and can therefore be written in the form

$$\Delta_{\{\lambda\}} \times \Delta_A = \Delta_{\{\tilde{\lambda}\}} \tag{17.17}$$

To find the representation $\Delta_{\{\tilde{\lambda}\}}$ let us suppose that the function $F_{\{\lambda\}}(y_1, y_2, \ldots, y_n)$ transforms in accordance with the representation $\Delta_{\{\lambda\}}$. According to (15.29), it can be regarded as being antisymmetric with respect to the interchange of the arguments whose numbers lie in the columns of the Young's tableau $\{\lambda\}$. Suppose that the function $F_A(y_1, \ldots, y_n)$ is antisymmetric with respect to the interchange of its arguments. The function $F_{\{\lambda\}}F_A$ must then belong to the basis of the irreducible representation $\Delta_{\{\tilde{\lambda}\}}$. It is clear that the function $F_{\{\lambda\}}F_A$ will be symmetric with respect to the interchange of the arguments whose numbers lie in the columns of the Young's tableau $\{\lambda\}$. It can be verified that when the operator $\Omega_{\{\lambda'\}}$ is applied to the function $F_{\{\lambda\}}F_A$, the result is not zero only for the operator whose Young's tableau is the transpose of $\{\lambda\}$. We thus conclude that the Young's tableau $\{\tilde{\lambda}\}$ should be the transpose of the tableau $\{\lambda\}$.

Consider now the two irreducible representations $\Delta_{\{\lambda\}}$ and $\Delta_{\{\lambda'\}}$ of the group S_n. Let us denote the characters of these representations by $\chi^{\{\lambda\}}(p)$ and $\chi^{\{\lambda'\}}(p)$, respectively. The character of the antisymmetric representation Δ_A is, of course, equal to $(-1)^{\varepsilon(p)}$. To find the number $r^{(A)}_{\{\lambda\}\{\lambda'\}}$ showing how many times the representation Δ_A is contained in the direct product $\Delta_{\{\lambda\}} \times \Delta_{\{\lambda'\}}$, let us use the formula given by (3.88)

$$r^{(A)}_{\{\lambda\}\{\lambda'\}} = \frac{1}{n!} \sum_p (-1)^{\varepsilon(p)} \chi^{\{\lambda\}}(p) \chi^{\{\lambda'\}}(p) \tag{17.18}$$

However,

$$(-1)^{\varepsilon(p)}\chi^{\{\lambda\}}(p) = \chi^{\{\tilde{\lambda}\}}(p) \tag{17.19}$$

and, consequently,

$$r^{(A)}_{\{\lambda\}\{\lambda'\}} = \frac{1}{n!}\sum_{p}\chi^{\{\tilde{\lambda}\}}(p)\chi^{\{\lambda'\}}(p) \tag{17.20}$$

From orthogonality of the characters of irreducible representations we find that $r^{(A)}_{\{\lambda\}\{\lambda'\}} = 1$ if $\{\lambda'\} = \{\tilde{\lambda}\}$, and $r^{(A)}_{\{\lambda\}\{\lambda'\}} = 0$ in all remaining cases. This proves the theorem.

The representation $p(e)^n$ can be decomposed into irreducible representations as shown in the previous section and the Young's tableaux for this consist of not more than two lines. In accordance with the above theorem, we can now conclude that the admissible irreducible representations $\Delta_{\{\lambda\}}$, which govern the transformation of the coordinate functions can have Young's tableaux of not more than two columns. The solutions of the Schroedinger equation (17.1), which transform in accordance with other representations of the permutation group, have no physical significance in our problem.

Let us now suppose that we know a solution $\Psi_{\{\lambda\}}(r_1, r_2, \ldots, r_n)$ of the Schroedinger equation (17.1), which transforms in accordance with the irreducible representation $\Delta_{\{\lambda\}}$ whose Young's tableau consists of not more than two columns. An arbitrary spin function $\chi(\sigma_1, \sigma_2, \ldots, \sigma_n)$ can be expanded in terms of the symmetrized spin functions $\chi_{\{\lambda\}}(\sigma_1, \ldots, \sigma_n)$ which transform in accordance with the irreducible representations $\Delta_{\{\lambda'\}}$:

$$\chi(\sigma_1, \sigma_2, \ldots, \sigma_n) = \sum_{\{\lambda'\}} c_{\{\lambda'\}}\chi_{\{\lambda'\}}(\sigma_1, \ldots, \sigma_n) \tag{17.21}$$

It is clear that when the product $\Psi_{\{\lambda\}}(r_1, \ldots, r_n)\cdot\chi(\sigma_1, \ldots, \sigma_n)$ is antisymmetrized, a non-zero result is contributed only by the spin function which transforms in accordance with the irreducible representation of the transposed Young's tableau $\{\tilde{\lambda}\}$.

We know that all the components of the spin function $\chi_{\{\lambda'\}}(\sigma_1, \sigma_2, \ldots, \sigma_n)$ are the eigenfunctions of the square of the total spin:

$$\hat{S}^2\chi_{\{\lambda'\}}(\sigma_1, \ldots, \sigma_n) = \hbar^2\left(\frac{\lambda_1' - \lambda_2'}{2}\right)\left(\frac{\lambda_1' - \lambda_2'}{2} + 1\right)\chi_{\{\lambda\}}(\sigma_1, \ldots, \sigma_n) \quad (17.22)$$

Since the complete antisymmetric wave function is a linear combination of such functions, it will also be an eigenfunction of the operator $\hat{S}^2$ with the same eigenvalue.

Thus, the classification of the eigenvalues of the Schroedinger equation (17.1) in accordance with the irreducible representations of the permutation group has, in view of Pauli's principle, turned out to be equivalent to the classification in accordance with the eigenvalues of the square of the total spin.

17.4 Symmetry properties of the coordinate wave function

The coordinate wave function which is the solution of the Schroedinger equation (17.1) should transform in accordance with an irreducible representation of the permutation group for which the Young's tableau consists of two columns. The function $\Psi_{\{\lambda\}}(\mathbf{r}_1, \mathbf{r}_2, \ldots, \mathbf{r}_n)$, which has this symmetry property, can be constructed from an arbitrary function $\psi(\mathbf{r}_1, \mathbf{r}_2, \ldots, \mathbf{r}_n)$ through symmetrization with respect to the numbers of the arguments lying in the rows of the Young's tableau, and subsequent antisymmetrization with respect to the numbers of the arguments in the columns:

1	$k+1$
2	$k+2$
⋮	⋮
⋮	n
k	

The function $\Psi_{\{\lambda\}}(r_1, r_2, \ldots, r_n)$ obtained in this way must be antisymmetric with respect to interchange within two groups of arguments – for example,

$$\left.\begin{aligned} &\Psi_{\{\lambda\}}(r_1, r_2, \ldots, r_k | r_{k+1}, \ldots, r_n) = \\ &\quad = -\Psi_{\{\lambda\}}(r_2, r_1, r_3, \ldots, r_k | r_{k+1}, \ldots, r_n) \\ &\Psi_{\{\lambda\}}(r_1, r_2, \ldots, r_k | r_{k+1}, \ldots, r_n) = \\ &= -\Psi_{\{\lambda\}}(r_1, r_2, \ldots, r_k | r_{k+2}, r_{k+1}, r_{k+3}, \ldots, r_n) \\ &\qquad\qquad k \geqslant n - k. \end{aligned}\right\} \tag{17.23}$$

It is clear that such a function cannot be antisymmetrized with respect to more than k arguments, since this would require antisymmetrization with respect to the arguments used initially to carry out the symmetrization operation. This property of $\Psi_{\{\lambda\}}$ can be written, for example, in the form of the following equation:

$$\Psi_{\{\lambda\}}(r_1, r_2, \ldots, r_k | r_{k+1}, \ldots, r_n)$$
$$= \sum_i \Psi_{\{\lambda\}}(r_1, \ldots, r_{i-1}, r_k, r_{i+1}, \ldots, r_k | r_i, r_{k+2}, \ldots, r_n) \tag{17.24}$$

The symmetry properties (17.23) and (17.24) of the coordinate wave function were first deduced by V. A. Fock (Zh. eksp. teor. Fiz., V 10, p. 388 (1950)).

In conclusion, consider the method of constructing the coordinate wave function corresponding to a given eigenvalue of $\hat{S}^2$ from independent one-electron functions $\varphi_1(r_1)$, $\varphi_2(r_2)$, $\ldots$, $\varphi_n(r_n)$.. If the eigenvalue of $\hat{S}^2$ is $\hbar^2 s(s+1)$, the irreducible representation by which the coordinate wave function must transform is determined by a Young's tableau containing two columns with lengths $\lambda_1' = \frac{n}{2} + s$, $\lambda_2' = \frac{n}{2} - s$. Let us fill the cells of this tableau with the numbers of the functions φ_i, and then construct the corresponding operator $\hat{\Omega}_{\{\lambda\}}$. If we apply this operator to the product of the one-electron functions, we obtain

$$\hat{\Omega}_{\{\lambda\}} \varphi_1(r_1) \ldots \varphi_n(r_n) = \sum_q \hat{Q} \sum_p (-1)^{\varepsilon(p)} \hat{P} \varphi_1(r_1) \varphi_2(r_2) \ldots \varphi_n(r_n) \tag{17.25}$$

Antisymmetrization applied to the columns of the tableau yields

$$\sum_p (-1)^{\varepsilon(p)} \hat{P} \varphi_1(\boldsymbol{r}_1) \ldots \varphi_n(\boldsymbol{r}_n)$$

$$= \begin{vmatrix} \varphi_1(\boldsymbol{r}_1) & \varphi_1(\boldsymbol{r}_2) & \ldots & \varphi_1(\boldsymbol{r}_{\lambda_1'}) \\ \varphi_2(\boldsymbol{r}_1) & \varphi_2(\boldsymbol{r}_2) & \ldots & \varphi_2(\boldsymbol{r}_{\lambda_1'}) \\ \ldots & \ldots & \ldots & \ldots \\ \varphi_{\lambda_1'}(\boldsymbol{r}_1) & \varphi_{\lambda_1'}(\boldsymbol{r}_2) & \ldots & \varphi_{\lambda_1'}(\boldsymbol{r}_{\lambda_1'}) \end{vmatrix} \times \begin{vmatrix} \varphi_{\lambda_1'+1}(\boldsymbol{r}_{\lambda_1'+1}) & \ldots & \varphi_{\lambda_1'+1}(\boldsymbol{r}_n) \\ \varphi_{\lambda_1'+2}(\boldsymbol{r}_{\lambda_1'+1}) & \ldots & \varphi_{\lambda_1'+2}(\boldsymbol{r}_n) \\ \ldots & \ldots & \ldots \\ \varphi_n(\boldsymbol{r}_{\lambda_1'+1}) & \ldots & \varphi_n(\boldsymbol{r}_n) \end{vmatrix} \qquad (17.26)$$

The resulting product of two determinants must then be symmetrized with respect to pairs of functions whose numbers lie in the rows of the table. The final expression for the multi-electron coordinate function will therefore be the sum of such products.

Exercises

17. 1. Given the six one-electron functions $\varphi_1(\boldsymbol{r}_1)$, $\varphi_2(\boldsymbol{r}_2)$, ..., $\varphi_6(\boldsymbol{r}_6)$, use Young's operators to construct coordinate wave functions for a set of six electrons corresponding to different possible eigenvalues of the square of the total spin.

17. 2. Show that if the one-electron coordinate wave functions for a system of an even number of electrons are identical in pairs, the corresponding value of the square of the total spin is zero.

Chapter 18

Symmetry Properties of Wave Functions for a System of Identical Particles with Arbitrary Spins

In this chapter we shall generalize the analysis given in Chapter 17 to the case of a system of identical particles with arbitrary spins. Identical nuclei in a polyatomic molecule are an example of such a system. The results obtained in this chapter will be used later for the classification of the states of multi-electron atoms.

18.1 Formulation of the problem

As before, we shall be dealing with the Schroedinger equation for a set of identical particles, which does not contain the spin operators. The particle spins will be regarded as arbitrary, i.e. not necessarily equal to one-half. This is, in fact, the first generalization of the previous analysis. Moreover, we shall suppose that, as a result of certain simplifications, the original Schroedinger equation is replaced by an approximate equation in which the symmetry with respect to the interchange of identical particles has been lost. For example, for a system of identical nuclei in the molecular problem (in the adiabatic approximation) we have the Schroedinger equation

$$\left\{-\frac{\hbar^2}{2M}\sum_{i=1}^{n}\Delta_i + U(\boldsymbol{R}_1, \ldots, \boldsymbol{R}_n)\right\}\Psi = E\Psi \tag{18.1}$$

which is invariant under the interchange of the variables $\boldsymbol{R}_i$. In the harmonic approximation, in which the potential energy $U(\boldsymbol{R}_1, \boldsymbol{R}_2, \ldots, \boldsymbol{R}_n)$ is a quadratic form involving small displacements from the positions of equilibrium, we have

$$\left\{-\frac{\hbar^2}{2M}\sum_{i=1}^{n}\Delta_i + \sum_{i,k=1}^{n}\left(\frac{\partial^2 U}{\partial \boldsymbol{R}_i \partial \boldsymbol{R}_k}\right)_{R=R^{(0)}}(\boldsymbol{R}_i - \boldsymbol{R}_i^{(0)})(\boldsymbol{R}_k - \boldsymbol{R}_k^{(0)})\right\}\Psi = 0 \tag{18.2}$$

This equation is valid in a small neighborhood of the point $\{\boldsymbol{R}_i^{(0)}\}$ of the configuration $\{\boldsymbol{R}_i\}$. However, if we recall that the rotational degrees of freedom, which describe the rotation of the molecule as a whole, are unaffected by the above approximation, we find that this equation is also valid in the neighborhood of any point $\{g\boldsymbol{R}_i^{(0)}\}$, where g is an arbitrary rotation. To derive the equation valid in some other region – for example, in the region of the point $\{\boldsymbol{R}_2^{(0)}, \boldsymbol{R}_1^{(0)}, \boldsymbol{R}_3^{(0)}, \ldots, \boldsymbol{R}_n^{(0)}\}$, where $\boldsymbol{R}_2^{(0)}$ and $\boldsymbol{R}_1^{(0)}$ are the positions of equilibrium of equivalent nuclei – we can simply carry out the corresponding interchange of the variables $\boldsymbol{R}_1$ and $\boldsymbol{R}_2$ in (18.2). It is clear that the solutions of the new equation can be obtained from the solution of (18.2) by the same permutation. The equation given by (18.2) is invariant for some interchanges of the variables $\boldsymbol{R}_i$ describing equivalent nuclei. In particular, it is invariant with respect to the three interchanges corresponding to rotations or mirror rotations which bring into coincidence the positions of equilibrium of equivalent nuclei. In fact, it is readily verified that this interchange is equivalent to rotation (or mirror rotation) followed by translations of the kind considered in Chapter 5 (see Fig. 1). We shall therefore assume that the eigenfunctions of the approximate equation transform in accordance with the representation of only a sub-group of the group S_n. This is one of the rare cases where transition to a more approximate model reduces the degree of symmetry rather than otherwise. This is, in fact, the second generalization.

Let us suppose that the (approximate) Schroedinger equation has been solved, i.e. we have found its eigenvalues and eigenfunctions. The important problem which arises therefrom is the determination of the statistical weights of each level. The statistical weight is defined as the number of independent states corresponding to the given level or, in other words, the number of symmetric (antisymmetric) functions which can be constructed from the given eigenfunctions and an arbitrary spin function. By an arbitrary spin function we mean a spin function having $(2s+1)^n$ independent components, where s is the maximum projection of the spin of a single particle and n is the number of particles in the system. We note, however, that when we consider the molecular problem, we must remember that the electron wave function will also have a definite symmetry with respect to the interchanges of $\boldsymbol{R}_i$ belonging to the above sub-group. These interchanges are equivalent to rotations or reflections of the electron coordinates. Therefore, when we determine the statistical weight of a level, we must consider the product $\Psi_{el}\,\Psi_{nucl}\,\chi_{nucl\ spin}$. Our analysis will refer to the case where Ψ_{el} transforms in accordance with the identity representation of the molecular symmetry group.

Let us now outline the plan of solution of the above problem: (1) We must determine the representation of the permutation group which corresponds to the solution of the approximate Schroedinger equation. (2) We then have to find the representation of the permutation group which governs the transformation of the $(2s+1)^n$ component spin function. (3) Finally, we set up the direct product of these representations and determine the number of times the symmetric (or antisymmetric) representation is contained in it. This number will, in fact, be equal to the statistical weight of the given level.

18.2 Frobenius' theorem

Let H denote the symmetry group of the approximate Schroedinger equation. The eigenfunctions of this equation corresponding to an eigenvalue form a manifold R_0 which is invariant with respect to a sub-group of H but not the entire group S_n. The results which we shall obtain in this section will be general and unconnected with the specific properties of the permutation group. Therefore, instead of the group S_n we shall consider an arbitrary group G of order N.

Let us decompose the group G into cosets of H:

$$H,\ g_1H,\ g_2H,\ \ldots,\ g_mH \tag{18.3}$$

The transformations $\hat{g}_1, \hat{g}_2, \ldots, \hat{g}_m$ will transform the space R_0 into new spaces R_i:

$$\hat{g}_i R_0 = R_i$$

The direct sum of the spaces $R = R_0 \oplus R_1 \oplus \ldots \oplus R_m$ is invariant with respect to the entire group G. In fact, let us take the arbitrary element from G – for example, $g = g_i h,\ h \in H$. We then have

$$\hat{g} R_j = \hat{g}_i \hat{h} \hat{g}_j R_0 = \hat{g}_k \hat{h}' R_0 \tag{18.4}$$

where $h' \in H$. Consequently,

$$\hat{g} R_j = \hat{g}_k R_0 = R_k \tag{18.5}$$

Let γ denote the representation of the group H which is realized in R_0, and let Γ denote the representation of G in the space R. The problem is to determine the characters of the representation Γ. Let us first find the characters $X(h)$ of Γ for $h \in H$. To do this, we shall first determine the number of sub-spaces R_j which are invariant under transformations from H. Since

$$\hat{h} R_j = \hat{h} \hat{g}_j R_0 = \hat{g}_j (\hat{g}_j^{-1} \hat{h} \hat{g}_j) R_0 \tag{18.6}$$

the invariance condition will be satisfied provided

$$g_j^{-1}hg_j \in H \tag{18.7}$$

If this is so, the transformation matrix in the space R_j corresponding to the element h will coincide with the representation matrix γ corresponding to the element $g_j^{-1}hg_j$. Since g_j does not belong to H, the element $g_j^{-1}hg_j$ may not belong to the same class C of the group H as the element h but, of course, it always belongs to the class K of G which contains C. Let us fix the element $h \in H$ and determine the number of sub-spaces R_j which transform in a similar way under the transformation h. To do this, let us establish how many elements $g_i^{-1}hg_i$ $(i = 0, \ldots, m)$ belong to the same class C of H. If for g_j the condition

$$g_j^{-1}hg_j \in C \tag{18.8}$$

is satisfied, we obtain the same result for all the elements of the coset $g_i\hat{H}$. Let us therefore first determine the number of elements $g^{-1}hg$, for g running over the entire group G, which belong to C and then divide the result by the order of the group H. If g runs over the entire group G, then the set $g^{-1}hg$ will contain each element of the class K just $\frac{N}{k}$ times, where k is the number of elements in the class K. If c is the number of elements in the class C, then the set under consideration will contain $N\frac{c}{k}$ elements belonging to this class. If we now allow for the fact that the element does not run over the entire group, but assumes only the values $g_0, g_1, \ldots, g_m$, we obtain the required number l_C showing how many elements in the set $g_j^{-1}hg_j$ belong to the class C which is contained in the class K. This is given by

$$l_C = \frac{1}{N_1}\frac{N}{k}c \tag{18.9}$$

where N_1 is the order of the sub-group H. If χ_C denotes the character of the representation of the class C of the sub-group

H, which is realized in the space R_0, then the character of the representation of the class K of G, which is realized in the space R, will be given by

$$X_K = \sum_{C \in K} l_C \chi_C = \frac{N}{N_1} \sum_{C \in K} \frac{c}{k} \chi_C \tag{18.10}$$

where the sum is evaluated over the classes C contained in K. The characters of the classes K' which are not represented in H are all zero, since transformations in these classes do not leave any of the sub-spaces R_i invariant. We shall show that the representation γ of the group H with characters χ_C induces a representation Γ of G with characters X_K. If the representation γ is irreducible, the representation Γ will, in general, be reducible. Its decomposition into irreducible representations is governed by the following theorem.

Frobenius' theorem

The representation Γ of the group G which is induced by the irreducible representation $\gamma^{(i)}$ of the sub-group H contains each irreducible representation $\Gamma^{\{\lambda\}}$ of G the same number of times as the representation of the group H, given by the matrices $\Gamma^{(\lambda)}$, is contained in $\gamma^{(i)}$.

The proof of this theorem is now quite simple. The numbers which we require are given by

$$r_{i\{\lambda\}} = \frac{1}{N} \sum_{K} k \bar{X}_K^{(i)} X_K^{\{\lambda\}} \tag{18.11}$$

where $X_K^{(i)}$ is the character of the representation Γ and $X_K^{\{\lambda\}}$ is the character of the irreducible representation $\Gamma^{\{\lambda\}}$. If we substitute (18.10) into the above equation, we obtain

$$r_{i\{\lambda\}} = \frac{1}{N} \sum_{K} k \frac{N}{N_1} X_K^{\{\lambda\}} \sum_{C \in K} \frac{c}{k} \bar{\chi}_C^{(i)} = \frac{1}{N_1} \sum_{C} c \bar{\chi}_C^{(i)} X_C^{\{\lambda\}} \tag{18.12}$$

which was to be proved.

Returning now to our original problem, we may conclude that the characters of the representation of the permutation group which transform the wave function under consideration are given by (18.10) and the decomposition of this representation into irreducible representations can be found using the Frobenius' theorem.

18.3 s-tensors

We can now take the next step and find the representation of the permutation group which governs the transformation of the $(2s+1)^n$ -component spin function. This function can be regarded as a tensor of rank n in a $2s+1$ -dimensional space, since rotations of the three-dimensional space transform this function so that

$$\chi'(\sigma_1, \sigma_2, \ldots, \sigma_n) = \sum_{\sigma'} D_{\sigma_1\sigma_1'} \ldots D_{\sigma_n\sigma_n'} \chi(\sigma_1', \ldots, \sigma_n') \quad (18.13)$$

where $\| D_{\sigma_i\sigma_i'} \|$ is the matrix of the irreducible representation of weight s of the rotation group. For the sake of brevity we shall refer to it as the s-tensor of rank n. It is clear that any s-tensor of rank n can be expanded in terms of $(2s+1)^n$ independent tensors forming the basis of the tensor space under consideration. In this space we can define a representation $p(E_{2s+1})^n$ of the permutation group (see Section 16.3). If we apply the permutation operator to the s-tensor, we obtain

$$\hat{p}F(\sigma_1, \sigma_2, \ldots, \sigma_n) = F(\sigma_{p_1}, \sigma_{p_2}, \ldots, \sigma_{p_n}) \quad (18.14)$$

An s-tensor of rank n can be expressed in terms of tensors which transform in accordance with irreducible representations of the permutation group. Such tensors can be obtained with the aid of Young's operators. It is clear that the admissible operators will be only those for which Young's tableaux contain not more than $2s+1$ rows. This is connected with the fact that the s-tensor which is antisymmetric with respect to more than $2s+1$ indices is zero. It may be shown that the multiplicity of the irreducible representations $\Delta_{\{\lambda\}}$ in the representation $p(E_{2s+1})^n$ is

$$\delta_{\{\lambda\}}=\frac{\prod\limits_{i>k}(l_i-l_k)}{(m-1)!\,(m-2)!\ldots 2!} \tag{18.15}$$

where $l_j=\lambda_j+m-1$, $m=2s+1$. The quantity $\delta_{\{\lambda\}}$ gives the number of independent components of the tensor which transforms in accordance with the irreducible representation $\Delta_{\{\lambda\}}$ of the permutation group (compare the discussion of spinors of rank n in Section 17.2).

18.4 Statistical weight of an energy level

Following these preliminary discussions we can go on to find the expression for the statistical weight f of a given energy level. We shall use the fact that the identity (symmetric) representation of the permutation group is contained only in the direct product of equivalent irreducible representations, while the antisymmetric representation is contained in the direct product of irreducible representations with transposed Young's tableaux (Chapter 16). Bearing this in mind, we obtain

$$\left.\begin{aligned} f&=\sum_{i,\{\lambda\}} k_i r_{i\,\{\lambda\}}\,\delta_{\{\lambda\}}, \quad s-\text{ integer}\\ f&=\sum_{i,\{\lambda\}} k_i r_{i\,\{\lambda\}}\,\delta_{\{\tilde{\lambda}\}}, \quad s-\text{ half-integer}\end{aligned}\right\} \tag{18.16}$$

where k_i is a number showing how many times the irreducible representation $\gamma^{(i)}$ of the group H is contained in the representation transforming the solutions of the approximate Schroedinger equation. The numbers $r_{i\,\{\lambda\}}$ and $\delta_{\{\lambda\}}$ are determined, respectively, by (18.12) and (18.15). The symbol $\{\tilde{\lambda}\}$ denotes the transposed Young's tableau.

It is interesting to note that although the definition of the statistical weight is connected with the permutation group, the statistical weights of levels can also be defined without bringing in the representations of the permutation group by considering only the point group H. In fact, we can proceed as follows. Since the spin functions form the representation basis for the permutation group, they therefore also form the representation basis for the point group H which is a sub-group of S_n. To find the character of this representation let us suppose that the permutation p has the cyclic structure $(\alpha_1\alpha_2 \ldots \alpha_n)$. Non-zero contributions to the character are provided only by those components of the spin function which remain invariant under the above permutation, i.e. they have the same indices for each cycle of the permutation p. It is clear that the number of such components is equal to the character of the representation and can be written in the form

$$(2s+1)^{\alpha_1+\alpha_2+\cdots+\alpha_n} \tag{18.17}$$

If we set up all the possible products of spin and coordinate functions, we obtain a basis Ω of the representation γ of the point group. This representation is the direct product of representations governing the transformation of spin and coordinate functions. Consider a representation of the point group H which is obtained from the antisymmetric (symmetric) representation Δ_A (Δ_S) of the group S_n by choosing the corresponding elements. We shall denote it by $\gamma^{(A)}$($\gamma^{(S)}$). We shall show that the statistical weight of the level under consideration is equal to the multiplicity of this representation in the representation γ. In fact, let us extend the space spanned by the basis Ω, as was done before, for the space invariant with respect to group S_n. The representation of S_n which is induced by the representation γ of H is realized in the expanded space. According to Frobenius' theorem, the representation Δ_A (Δ_S) can be induced only by the representation $\gamma^{(A)}$($\gamma^{(S)}$). The multiplicity of the representation Δ_A (Δ_S) in the expanded representation must therefore be equal to the multiplicity of γ. This proves the above statement.

18.5 Eigenvalues of the total spin operator

To conclude this chapter, let us consider the eigenvalues of the total spin of a system corresponding to a given energy state. We have seen that, for a multi-electron system, to each eigenvalue of the spinless Schroedinger equation there corresponds a definite eigenvalue of the total spin. In the general case which we are considering, this one-to-one correspondence is not valid. To each energy level there will, in general, correspond a number of eigenvalues of the total spin. This is connected firstly with the fact that the s-tensor of rank n corresponding to the irreducible representation of the permutation group will now transform in accordance with a reducible representation of the rotation group. (Irreducibility occurs only for $\frac{1}{2}$-tensors or spinors.) Secondly, since the Schroedinger equation has the symmetry of a point group, it follows that its solution transforms, in general, in accordance with an irreducible representation of a permutation group. This means that the non-zero contribution to the complete antisymmetric (or symmetric) function is due to s-tensors transforming in accordance with irreducible representations of the group S_n. Therefore, even when $s = \frac{1}{2}$, a given energy level may correspond to states with different total spins.

Unfortunately, the problem of the representation of the rotation group which is realized on the components of an s-tensor with given symmetry with respect to permutation, cannot be outlined in a compact way.

Let $R^{\{\lambda\}}(D^{(s)})$ be the representation of the rotation group which governs the transformation of independent components of the s-tensor corresponding to an irreducible representation $\Delta_{\{\lambda\}}$ of the permutation group. To find the eigenvalues of the square of the total spin we must decompose the representation $R^{\{\lambda\}}(D^{(s)})$ into irreducible parts. The characters of this representation can be found from (16.28), but this requires a knowledge of the characters of irreducible representations of the group S_n. We shall confine our attention to quoting without proof a method which will enable us to find the decomposition of the representation $R^{\{\lambda\}}(D^{(s)})$ if this decomposition is known for the representations $R^n(D^{(s)})$ corresponding to the symmetric representation of the group S_n. Let $\chi^{(s,\,n)}$ be the character of the representation $R^n(D^{(s)})$. It can be found from (16.30), for example. Next, let us suppose that $\chi^{(s,\,0)} = 1$ and $\chi^{(s,\,k)} = 0$ if $k < 0$. The character $\chi^{(s,\,\{\lambda\})}$ of the representation $R^{\{\lambda\}}(D^{(s)})$ is equal to the following determinant:

$$\chi^{(s,\,\{\lambda\})} = \chi^{(s;\,\lambda_1,\,\lambda_2,\,\ldots,\,\lambda_k)} = \begin{vmatrix} \chi^{(s,\,\lambda_1)}\chi^{(s,\,\lambda_1+1)} & \cdots & \cdots \\ \chi^{(s,\,\lambda_2-1)}\chi^{(s,\,\lambda_2)}\chi^{(s,\,\lambda_2+1)} & \cdots & \cdots \\ \cdots & \cdots & \cdots \\ \cdots & \cdots & \chi^{(s,\,\lambda_k-1)}\chi^{(s,\,\lambda_k)} \end{vmatrix} \qquad (18.18)$$

Since we are interested in the decomposition of the representation $R^{\{\lambda\}}(D^{(s)})$ into irreducible representations, we can use the Clebsch-Gordan rule.

Exercises

18.1. Find the statistical weights of the vibrational levels of an octahedral molecule YX_6 for the ground and first excited states. Consider the case when the spin of the nucleus of the atom X is

$$\text{a) } s = \frac{1}{2}, \quad \text{b) } s = 1$$

18.2. Find the representation of the rotation group governing the transformation of a symmetric 1-tensor of rank n.

18.3. Find the representation of the rotation group which governs the transformation of the 1-tensor of rank 4 with different Young's tableaux.

Chapter 19

Classification of the States of a Multi-Electron Atom

In Chapter 17 we discussed the symmetry properties of a multielectron wave function which was the eigenfunction of the Hamiltonian without the spin operators. The only symmetry property of this Hamiltonian which we have used was invariance under the interchange of the electron coordinates. We have seen that, as a result of this invariance and of Pauli's principle, the states of a multi-electron system can be classified in accordance with the eigenvalues of the square of the total spin.

We shall now consider the classification of the states of a multi-electron atom. We shall introduce a number of additional assumptions with regard to the symmetry of the Hamiltonian and then, by a generalization of the spinless Hamiltonian, we shall take into account the spin-orbital interaction.

19.1 Configuration

As a first approximation, the interaction between electrons in an atom can be replaced by an effective spherically symmetric field. Each electron can then be regarded as moving independently in this field and in the field of the nucleus. The result of this approximation is the non-interacting electron model, which, as we shall see, has the maximum symmetry. We shall also neglect, at least to begin with, the spin-orbital interaction. Since the potential energy in this approximation has spherical symmetry, the single-electron states, as we know, are classifiable in accordance with the irreducible representations of the three-dimensional rotation group, i.e. they can be classified in terms of the azimuthal quantum number l (see Chapter 13).

To distinguish different levels with the same l-values, we must introduce an additional quantum number n which is the analog of the principal quantum number of the hydrogen atom. The values of the principal quantum number can, of course, be chosen arbitrarily. They are in fact usually chosen so that there is a correspondence with the classification of states in the Coulomb field. Each energy level E_{nl} is degenerate with respect to a component of the angular momentum in a fixed direction, i.e. with respect to the quantum number m. The single-electron coordinate wave functions can therefore be denoted by $\psi_{nlm}(\boldsymbol{r})$. If the atom contains N electrons, its state will be determined by a set of $3N$ single-electron quantum numbers n_i, l_i, m_i. The set of quantum numbers n_i, l_i which in our model define the energy of the atom, namely

$$E = \sum_{i=1}^{N} E_{n_i l_i} \tag{19.1}$$

is called the configuration of the atom. It is customary in spectroscopy to use the following notation. Single-electron states with quantum numbers $l = 0, 1, 2, \ldots$ are represented by the letters $s, p, d, \ldots$, respectively. These symbols are preceded by the value of the principal quantum number, and the number of electrons with given numbers n and l is indicated by an exponent. For example, the configuration $(1s)^2(2s)^2(2p)^6 3s$ consists of two electrons in the state with $n=1$, $l=0$, two electrons in the state with $n=2$, $l=0$, six electrons with $n=2$, $l=1$, and one electron in the state with $n=3$, $l=0$. To each configuration we can assign the parity quantum number $w=\pm 1$, which determines the behavior of the wave function under inversion. We know (see Chapter 13) that for a single electron $w=(-1)^l$. For a multi-electron system, therefore, we have

$$w=(-1)^{\sum_i l_i}$$

However, even in the non-interacting electrons model we cannot completely neglect the mutual effects electrons since we must take into account Pauli's principle which prevents more than two electrons occupying a given state (with the spin ignored). For each configuration allowed by Pauli's principle we can construct, in general, a number of antisymmetric total wave functions with different eigenvalues of the square of the total spin. There is a definite correspondence between the configuration and the eigenvalue of the square of the total spin which we shall discuss in detail when we consider a more realistic model of the atom. We conclude our discussion of the non-interacting electron model by noting its symmetry group. This group can clearly be represented in the form of the direct product N of three-dimensional rotation groups (or, more precisely, orthogonal groups) and the permutation group for the space variables:

$$G=O(3)\times O(3)\times \dots \times O(3)\times S_n \tag{19.2}$$

19.2 Spectral terms

We must now take directly into account the interaction

between the electrons. The presence of terms of the form $\frac{1}{|r_i - r_k|}$ in the energy operator ensures that the Hamiltonian will not be invariant under independent rotations of the position vectors r_i of the individual electrons. The Hamiltonian will be invariant only when all the r_i are rotated in the same way. Therefore, the symmetry group of this model can be represented by

$$G = O(3) \times S_n \tag{19.3}$$

The classification of the energy states of an atom in accordance with the irreducible representations of the group $O(3)$ corresponds to the classification in accordance with the eigenvalues of the square of the total orbital angular momentum, or the quantum number L, and the parity quantum number w. We note that, for a spectral term, the parity quantum number will no longer be determined by the orbital quantum number as in the case of a single electron. Classification in accordance with the irreducible representations of the permutation group S_n is equivalent, as we know, to the classification in accordance with the eigenvalues of the square of the total spin or quantum number S. The degree of degeneracy of an energy level E_{LS} in this model is $(2L+1)(2S+1)$. The set of degenerate states corresponding to given values of L and S is usually referred to as a spectral term. These terms are denoted by capital letters S, P, D, ... with a left-hand superscript equal to the multiplicity $2S+1$. For example, the term $\left(L=1,\ S=\frac{1}{2}\right)$ is denoted by ${}^2P$.

One of the most effective and widely used methods of solving multi-electron problems is the Hartree-Fock method. It is based on the variational principle for the energy. The wave functions involved are constructed from single-electron functions, the latter being taken in the form

$$\psi_{nlm}(r) = R_{nl}(r)\, Y_l^{(m)}(\theta,\ \varphi) \tag{19.4}$$

where $Y_l^{(m)}(\theta,\ \varphi)$ are the spherical harmonics. For the functions $R_{nl}(r)$ we then have a set of integral differential equations which is solved by the method of successive approximations. We need not for the moment consider in detail this method of solving the multi-electron problem. All that is important at present is that the electron functions are chosen in the form given by (19.4) and, consequently, the complete wave function is constructed from single-electron functions belonging to a definite configuration, i.e. a definite set of quantum numbers n_i, l_i, m_i. The single-electron states with the same values of n, l form an electron shell. If the electrons occupy all the states in a given shell (with allowance for the spin states) the shell is said to be filled. Interaction between electrons in a given shell, which have the same radial function, is stronger than the interaction between electrons belonging to different shells. The latter can be taken into account through an effective screening of the field of the nucleus. In this approximation, the problem reduces to finding the wave functions for the individual shells. On the other hand, we have seen that the total wave function is characterized by the quantum numbers L and S. It is therefore interesting to establish the spectral terms corresponding to a given configuration of single-electron states. We shall consider this problem for a configuration describing a given shell.

19.3 Correspondence between configurations and terms

Suppose that there are k electrons in a shell l, and consider the product of the single-electron coordinate functions of these electrons (for the sake of brevity we shall omit the indices n, l):

$$\psi_{m_1}(\mathbf{r}_1)\,\psi_{m_2}(\mathbf{r}_2)\,\ldots\,\psi_{m_k}(\mathbf{r}_k) \tag{19.5}$$

If we substitute

$$\mathbf{r}_i \to g^{-1}\mathbf{r}_i \tag{19.6}$$

where g is an arbitrary transformation in the rotation group, each of the single-electron functions will transform in accordance with the rule

$$\psi_m(g^{-1}r) = \sum_{m'=-l}^{l} D^{(l)}_{m'm}(g)\,\psi_{m'}(r) \tag{19.7}$$

The matrix $\| D^{(l)}_{m'm}(g) \|$ is the matrix of the irreducible representation of weight l of the rotation group. It follows from (19.7) that, for the product of the single-electron functions (19.5), we have the following transformation rule:

$$\psi_{m_1}(g^{-1}r_1)\psi_{m_2}(g^{-1}r_2)\ldots\psi_{m_k}(g^{-1}r_k) = \sum_{m'} D^{(l)}_{m_1'm_1}(g)\,D^{(l)}_{m_2'm_2}(g)$$

$$\ldots D^{(l)}_{m_k'm_k}(g) \times \psi_{m_1'}(r_1)\,\psi_{m_2'}(r_2)\ldots\psi_{m_k'}(r_k) \tag{19.8}$$

We thus see that, for transformations in the rotation group, the product of the single-electron functions corresponding to a given shell transforms as an l-tensor. We note that the coordinate wave function of a multi-electron system should transform under the interchange of the arguments in accordance with one of the irreducible representations of the permutation group for which the Young's tableau must consist of not more than two columns. The construction of the Schroedinger wave function is therefore reduced to the decomposition of an arbitrary l-tensor of rank k into symmetrized tensors which transform in accordance with irreducible representations of the permutation group with allowed Young's tableaux (consisting of two columns). This decomposition was discussed in Chapter 18. Independent components of the symmetrized l-tensor transform, as we already know, in accordance with the representation $R^{\{\lambda\}}(D^{(l)})$ of the rotation group.

To answer the question posed at the end of the last section, we must establish the irreducible representations of the rotation group into which the representation $R^{\{\lambda\}}(D^{(l)})$ can be decomposed. A method of solving this problem was considered in Chapter 18. This method can be used to find the value of the total orbital angular momentum.

As an illustration, consider the table of possible eigenvalues of the total orbital angular momentum for the configuration (p^k). Symmetry with respect to interchanges of the indices of the 1-tensor will be indicated by splitting into integral components.

Table of possible eigenvalues of the total orbital angular momentum for the configuration $(p)^k$

k	$\{\lambda\}$	L
1	1	P
2	2	SD
	1+1	P
3	2+1	PD
	1+1+1	S
4	2+2	SD
	2+1+1	P
5	2+2+1	P
6	2+2+2	S

It is clear from the above table that a given total angular momentum corresponds to a certain set of Young's tableaux. These Young's tableaux are obtained from one another by deleting the column containing the maximum number $(2l+1)$ cells (in our case (3) or by augmenting the Young's tableau up to 'maximum' – for example,

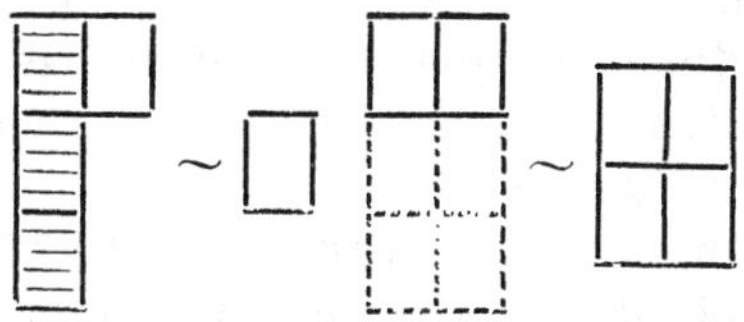

This is a general result but we shall not prove it here.

Since the symmetry of the spin function is determined by the transposed Young's tableau, it follows that, for the states given in the tableau, we can readily determine the eigenvalues of the total spin. Consider, for example, the configuration $(p)^4$. For the tableau $\{2, 2\}$ the transposed tableau will also be given by $\{2, 2\}$. Hence, the possible value of the total spin is $S=\frac{2-2}{2}=0$. For $\{2, 1, 1\}$ we have the transposed tableau $\{3, 1\}$, and $S=\frac{3-1}{2}=1$. The following terms are therefore possible for the configuration $(p)^4$: ^1S, ^1D, ^3P .

19.4 The spin-orbital interaction

Let us now consider a model of the atom which takes into account the spin-orbital interaction:

$$\sum \xi_i(r_i)\,\hat{s}_i\hat{l}_i \tag{19.9}$$

where $\hat{s}_i$ and $\hat{l}_i$ are the spin and orbital angular momentum operators of the i-th electron, and ξ_i are certain functions of r_i (see Chapter 13). It is clear that the operator (19. 9) does not commute separately with the operators $\hat{S}=\sum \hat{s}_i$ and $\hat{L}=\sum \hat{l}_i$. It is readily verified that it commutes only with the sum $(\hat{S}+\hat{L}^2)$ (we find that the spin-orbital interaction operator (19. 9) commutes with each of the operators $\hat{s}_i+\hat{l}_i$). This means that while so far we have been able to speak of the individual invariance under rotations of the arguments of the coordinate function and of induced rotations of the spin functions, all that remains now is invariance under the simultaneous application of the transformations. The energy levels will now be classified not in accordance with the quantum numbers L and S, but in accordance with the eigenvalues of the total angular momentum operator for the atom

$$J=L+S=\sum_i (s_i+l_i) \tag{19.10}$$

If the spin-orbital interaction can be regarded as a perturbation ($L-S$ coupling), the corresponding values of the total angular momentum for levels into which a given term is found to split can be obtained from the Clebsch-Gordan rule.

Consider the configuration $(p)^4$ as an example. The possible value of J for this configuration are:

Term	J	Level
^1S	0	$^1S_0$
^1D	2	$^1D_2$
^3P	0, 1, 2	$^3P_0$, $^3P_1$, $^3P_2$

In heavy atoms the spin-orbital interaction is strong and cannot be regarded as a perturbation. If it is greater than the interaction between the electrons, it must be the first to be taken into account. The interaction between the electrons can be partly taken into account through an effective spherically symmetric potential. The magnitude of the spin-orbital interaction must therefore be compared only with that part of the inter-electron interaction which does not enter into this effective potential.

In the zeroth approximation each electron can be considered individually, and the single-electron states can be characterized by the eigenvalues of the total single-electron angular momentum operator $\hat{j} = \hat{l}_i + \hat{s}_i$. We must then take into account the interaction between the electrons, and this means that the energy levels of the complete atom will again be classified by the eigenvalues of the total angular momentum. This approximation is called the $j-j$ coupling. If the spin-orbital interaction is strong enough, so that the single-electron levels with different j are highly split, we can introduce the concept of a j shell which is the analog of the l-shell. Since the multi-electron function must be antisymmetric under the interchange of the electrons, the wave function for the j-shell will be an antisymmetric j-tensor. To find the possible eigenfunctions of the total angular momentum we must decompose into irreducible parts the representation of the rotation group which is realized on the components of this tensor. This problem can be resolved in the same way as for the l-shell.

In practice, however, there is no need to carry out the relatively complex procedure each time because the results are well known and have been tabulated. As an illustration we reproduce below a table of the total angular moments for a number of j-shells. The number of electrons in a j-shell is equal to $2j+1$, although in the table the maximum value for a j-shell is $\frac{2j+1}{2}$. This is so because the classification of the states in the j-shell with k electrons is the same as for $2j+1-k$ electrons, in view of the rule given in Section 19.3.

Table of states for the j-shells

j	No. of electrons	J
$j=\frac{3}{2}$	1	3/2
	2	2, 0
$j=\frac{5}{2}$	1	5/2
	2	4, 2, 0
	3	9/2, 5/2, 3/2
$j=\frac{7}{2}$	1	7/2
	2	6, 4, 2, 0
	3	15/2, 11/2, 9/2, 7/2, 5/2, 3/2
$j=\frac{9}{2}$	1	9/2
	2	8, 6, 4, 2, 0
	3	21/2, 17/2, 15/2, 13/2, 11/2, $(9/2)^2$, 7/2, 5/2, 3/2
	4	12, 10, 9, $(8)^2$, 7, $(6)^3$, 5, $(4)^3$, 3, $(2)^2$, $(0)^2$
	5	25/2, 21/2, 19/2, $(17/2)^2$, $(15/2)^2$, $(13/2)^2$, $(11/2)^2$, $(9/2)^3$, $(7/2)^2$, $(5/2)^2$, 3/2, 1/2

Chapter 20

Applications of Group Theory To Problems Connected With the Perturbation Theory

The Schroedinger equation which determines the stationary states of a quantum-mechanical system can be solved exactly only in exceptional cases. One of the important methods for the approximate solution of the Schroedinger equation is the method of the theory of perturbations. This method can be used whenever the Hamiltonian can be written in the form

$$\hat{H} = \hat{H}_0 + \hat{V} \tag{20.1}$$

where $\hat{H}_0$ is the Hamiltonian of a problem for which a simple solution can be obtained and $\hat{V}$ is a perturbation operator. When the symmetry of the unperturbed operator $\hat{H}_0$ and the perturbation operator $\hat{V}$ are taken into account, the application of perturbation theory to specific problems is frequently much easier.

20.1 Splitting of energy levels under the action of a perturbation

Suppose that the Hamiltonian $\hat{H}_0$ has the symmetry group G_0 while the perturbation operator $\hat{V}$ has the symmetry group

G_1. Usually, the requirement that the operator H_0 must be relatively simple in comparison with the complete Hamiltonian $\hat{H} = \hat{H}_0 + \hat{V}$ implies a higher symmetry of $\hat{H}_0$ in comparison with the symmetry of $\hat{H}$. We shall consider two cases: (a) the group G_1 coincides with G_0; (b) G_1 is a sub-group of G_0.

a. The symmetry of the complete Hamiltonian is the same as the symmetry of the unperturbed problem: $G_0 = G_1$. We know that the eigenvalues of the Schroedinger equation can be classified by the irreducible representations of the symmetry group. Consequently, the classification and degree of degeneracy of the energy levels in our problem remain the same as in the unperturbed case. We can expect only a shift of the eigenvalues E_s of $\hat{H}$ relative to the eigenvalues $E_s^{(0)}$ of the operator $\hat{H}_0$:

$$E_s = E_s^{(0)} + \Delta E_s \tag{20.2}$$

Therefore, in this case, the perturbation cannot give rise to splitting of degenerate energy levels. The only exception to this rule occurs in the case of accidental degeneracy when the eigenfunctions corresponding to a given energy level of the unperturbed problem transform by a reducible representation of G_0.

If we vary the perturbation V but retain its symmetry, the eigenvalue shifts ΔE_s will also vary and some energy levels may intersect. Accidental degeneracy will occur at the points of intersection, since the eigenfunctions corresponding to this energy value will transform by a reducible representation of G_0. There is, however, a rule which, in some cases, forbids the intersection of levels corresponding to equivalent irreducible representations. Let us consider for the sake of simplicity two non-degenerate levels, assuming that the corresponding wave functions ψ_1 and ψ_2 transform by equivalent irreducible representations of G_0. We shall suppose that, for a certain value of the perturbation V_1, the energies E_1 and E_2 are almost equal. We shall try to establish whether the departure of the perturbation V from the value V_1 will give rise to the intersection of these levels. If we denote the difference $V - V_1$ by V', we can find the energy levels by solving the secular equation

$$\begin{vmatrix} E_1 + v'_{11} - E & v'_{12} \\ v'_{21} & E_2 + v'_{22} - E \end{vmatrix} = 0 \qquad (20.3)$$

We have

$$E = \frac{1}{2}(E_1 + E_2 + v'_{11} + v'_{22})$$

$$\pm \sqrt{(v'_{11} - v'_{22} + E_1 - E_2)^2 + 4\,|v'_{12}|^2} \qquad (20.4)$$

To ensure that the roots of the secular equation coincide, we must satisfy simultaneously the following two conditions

$$E_1 - E_2 + v'_{11} - v'_{22} = 0, \quad v'_{12} = 0$$

These conditions impose a relatively stringent restriction on the perturbation. If, for example, the perturbation is determined by only one parameter then, in general, we cannot simultaneously satisfy these two conditions and, consequently, level intersection is impossible. If the wave functions corresponding to these levels transform by inequivalent irreducible representations, the second condition $v'_{12} = 0$ is satisfied identically [cf. Equation (5. 32)] and level intersection can occur even when the perturbation is a function of only one parameter.

b. Consider now the case where the group G_1 is a sub-group of G_0. The eigenvalues of the Hamiltonian $\hat{H}$ must all be classified by irreducible representations of G_1. Since the orders of irreducible representations of a sub-group do not exceed the orders of the irreducible representations of the group, splitting of the unperturbed levels can occur in this case. When we speak of level splitting we implicitly assume that the perturbation is so small that the levels of the perturbed problem can be uniquely associated with the eigenvalues of $\hat{H}_0$.

Consider an eigenvalue $E^{(0)}$ of $\hat{H}_0$. Suppose that the corresponding eigenfunctions $\psi_1, \psi_2, \ldots, \psi_k$ transform by the irreducible representation $\Gamma(g)$ for operations g from G_0. Let E_l be the energy levels of the perturbed problem into which $E^{(0)}$ will split. The eigenfunctions $\psi_\alpha^{(l)}$ of $\hat{H}$, which correspond to each of the eigenvalues E_l, will transform by an irreducible representation γ_l of the group G_1 under operation from G. Let $\gamma = \sum_l^{\oplus} \gamma_l$ and let us reduce the perturbation $\hat{V}$ without limit and without change in its symmetry. For any value of $\hat{V}$ we can assume that the wave functions belonging to all the levels E_l will transform by a representation γ of G_1. In the limit $\hat{V} = 0$, we obtain the eigenfunctions of the unperturbed operator which are related to the functions $\psi_1, \psi_2, \ldots, \psi_k$ by a unitary transformation and, therefore, the representation $\Gamma(g)$ must be equivalent to the representation $\gamma(g)$ for $g \in G_1$:

$$\gamma(g) = U^{-1}\Gamma(g)U, \quad g \in G_1$$

and, consequently,

$$U^{-1}\Gamma(g)U = \sum_l^{\oplus} \gamma_l(g), \quad g \in G_1 \tag{20.5}$$

This result can be used to determine the number of components into which a given energy level $E^{(0)}$ will split when the perturbation is introduced. It is clear that all that is required is to decompose the representation $\Gamma(g)$ into irreducible representations of G_1.

We see that the introduction of a perturbation with lower symmetry than the symmetry of the initial problem leads to a partial removal of degeneracy. The degeneracy of each of the new levels is determined by the order of the corresponding irreducible representation of G_1. The above results are conveniently illustrated by the following scheme:

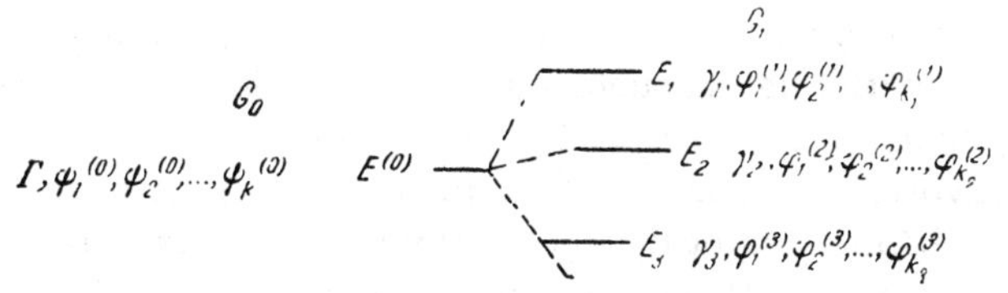

20.2 The zero-order approximation

Consider the degenerate level $E^{(0)}$ of the unperturbed problem. The orthonormal wave functions belonging to this level will be denoted by $\psi_1\ \psi_2, \ldots, \psi_k$. It is well known that the first-order perturbation-theory corrections ΔE to the energy are given by the secular equation

$$\begin{vmatrix} v_{11}-\Delta E & v_{12} & \ldots & v_{1k} \\ v_{21} & v_{22}-\Delta E & \ldots & v_{2k} \\ \ldots & \ldots & \ldots & \ldots \\ v_{k1} & v_{k2} & \ldots & v_{kk}-\Delta E \end{vmatrix} = 0 \qquad (20.6)$$

where

$$v_{ik} = \int \bar{\psi}_i V \psi_k \, d\tau. \qquad (20.7)$$

Linear combinations of the functions $\psi_1, \psi_2, \ldots, \psi_k$ for which the perturbation matrix is diagonal are called regular functions of the zero-order approximation. The eigenfunctions of the perturbed operator transform continuously from these functions when the perturbation is introduced. Since the perturbation operator V is invariant under the group G_1, its regular functions should transform (in the zero-order approximation) by the irreducible representations of this group (see Chapter 5). If in the decomposition of the representation Γ which transforms the functions $\psi_1, \psi_2, \ldots, \psi_k$ each irreducible representation of G_1 is encountered not more than once, then by constructing linear combinations of $\psi_1, \psi_2, \ldots, \psi_k$ which transform by the irreducible representations of G_1 we obtain the regular functions of the zero-order approximation. If, on the other hand, the same representation is encountered r times in the decomposition of Γ, then the diagonalization of the perturbation matrix will involve the solution of a secular equation of order r. The multiplicities of the irreducible representations are usually low and, therefore, the construction of functions transforming by the irreducible representations of G_1 makes the solution of (20. 6) much easier.

20.3 Atoms in inhomogeneous magnetic and electric fields

To illustrate the general theory developed in the previous sections, let us consider the splitting of the energy levels of an atom in an external homogeneous magnetic or electric field. We shall assume, for the sake of simplicity, that the state of the atom is determined by the state of one valence electron moving in the spherically symmetric field of the remainder of the atom. To explain the characteristic properties of these phenomena, we shall use a simplified theory which does not take spin into account.

a. Zeeman effect

We shall suppose that the atom is placed in a uniform magnetic field parallel to the z-axis. The operator representing the interaction between the electron and the magnetic field H can be written in the form

$$\hat{V} = \frac{e}{2mc} H\hat{l}_z \qquad (20.8)$$

where $\hat{l}_z$ is the z-component of the orbital angular momentum operator. If the symmetry group G_0 for the problem were to be the group $O(3)$, then the introduction of the interaction (20. 8) would ensure that the only remaining symmetry operations will be rotations about the z-axis and reflections in the xy-plane. The group G_1 will therefore coincide with the group $\mathbf{C}_{\infty h}$. We note that $\mathbf{C}_{\infty h}$ is Abelian and, consequently, all its irreducible representations are one-dimensional. The states in the unperturbed problem can be classified in accordance with the irreducible representations of the rotation group, i.e. the azimuthal quantum number l. Each level E_l is degenerate and the degree of degeneracy is $2l+1$. To determine the splitting of this level in the magnetic field we must decompose the representation $D^{(l)}$ of the rotation group in irreducible representations of $\mathbf{C}_{\infty h}$. We obtain this representation if we start with degenerate wave functions forming a canonical basis of an irreducible representation of the rotation group. In fact, in this case, the representation matrices corresponding to rotations about the z-axis have the diagonal form

$$D^{(l)}(0,\,0,\,\varphi)=\begin{pmatrix} e^{-il\varphi} & 0 & \dots & 0 \\ 0 & e^{-i(l-1)\varphi} & \dots & 0 \\ \cdot & \cdot & \cdot & \cdot \\ 0 & 0 & \dots & e^{il\varphi} \end{pmatrix} \tag{20.9}$$

The representation $D^{(l)}$ of the rotation group therefore decomposes into $2l+1$ irreducible representations $e^{im\varphi}$ ($m= -l,\,-l+1,\,\dots,\,l$) of the group $\mathbf{C}_{\infty h}$. We may therefore conclude that the energy levels E_l split into $2l+1$ sub-levels in the magnetic field, and each of these levels has a definite value of the magnetic quantum number: $m=-l,\,-l+1,\dots,\,l$. To each of these levels we can also assign the parity quantum number $w=(-1)^l$, since the operation of inversion remains a symmetry operation for the perturbed operator.

We note that in the above case the perturbation operator differs by only a constant factor from the infinitesimal operator of the symmetry group of the unperturbed operator $\hat{H}_0$ and, therefore, zero-order perturbation theory gives the exact solution of the problem.

b. Stark effect

Let us now suppose that the atom is placed in a uniform electric field parallel to the z-axis. The interaction between an electron and the electric field $\mathscr{E} = \mathscr{E}\mathbf{e}_z$ is now of the form

$$V = e\mathscr{E}z \tag{20.10}$$

In this case, the operator is also invariant under rotations about the z-axis. However, the symmetry under reflections of the position vector in the xy-plane is now lost, and instead we have symmetry under inversion i, which can be represented as a reflection in the xy-plane followed by 180° rotation about the z-axis. The introduction of an electric field, therefore, means that m is no longer a good quantum number. In addition to invariance under rotations about the z axis, the perturbation (20.10) is also invariant under reflections in planes containing this axis. The symmetry group of the perturbation will, in this case, be the group $\mathbf{C}_{\infty v}$.

We know (see Exercise 10. 1) that the group $\mathbf{C}_{\infty v}$ has two one-dimensional representations, and an infinite number of two-dimensional representations. The matrices of these representations are given below:

Group element	A_1	A_2	E_m
$C(\varphi)$	1	1	$\begin{pmatrix} e^{-im\varphi} & 0 \\ 0 & e^{im\varphi} \end{pmatrix}$
σ_v	1	-1	$\begin{pmatrix} 0 & 1 \\ 1 & 0 \end{pmatrix}$

To determine the splitting of a level E_l in the electric field, we must decompose the representation of $\mathbf{C}_{\infty v}$ given by the matrices $D^{(l)}$ into irreducible representations. Using

(20. 9), we have

$$\operatorname{Sp} D^{(l)}(0,\ 0,\ \varphi)=\sum_{k=1}^{l} 2\cos k\varphi+1=\sum_{k=1}^{l}\operatorname{Sp} E_k(\varphi)+1 \quad (20.11)$$

The character of the representations $D^{(l)}$ corresponding to an operation σ_v, for example σ_{zy}, can be evaluated as follows:

$$\operatorname{Sp} D^{(l)}(\sigma_{yz})=\operatorname{Sp} D^{(l)}(C_x(\pi)\times i)=\operatorname{Sp} D^{(l)}(C_x(\pi))\operatorname{Sp} D^{(l)}(i)$$
$$=(1+2\cos\pi+\ \ldots\ +2\cos l\pi)w=(-1)^l w$$

However, in the present case $w=(-1)^l$ and, therefore,

$$\operatorname{Sp} D^{(l)}(\sigma_v)=1 \quad (20.12)$$

Comparison of (20. 11) and (20. 12) with the characters of the irreducible representations of $C_{\infty v}$ shows that

$$D^{(l)}(g)=\sum_{k=1}^{l}{}^{\oplus} E_k(g)\oplus A_1,\quad g\in C_{\infty v} \quad (20.13)$$

Therefore, we may conclude that in a uniform electric field the level E_l splits into l doubly degenerate levels corresponding to the representations $E_k (k=1,\ \ldots,\ l)$, and one non-degenerate level whose wave function transforms in accordance with the representation $A^{(1)}$.

We note that the matrix elements of the perturbation operator for functions with the same azimuthal number l are zero because the product of two such functions is an even function, and the perturbation changes sign under the inversion transformation. As a result, the first-order perturbation-theory correction to the energy is zero. The splitting appears only in the second order and, therefore, its magnitude is proportional to the square of the field strength. The only exception is the hydrogen atom, for which the first-order correction is non-zero because of the additional degeneracy of its energy levels.

20.4 Atoms in a crystal field

Consider now the splitting of the energy levels of an atom in the field of a crystal. We shall suppose that the effect of the crystal on the atom can be looked upon as a small perturbation whose symmetry is determined by the symmetry of the crystal. Thus, for the group G_1, which should be a subgroup of the rotation group, we have in this case one of the point groups. Since the characters of the irreducible representations of point groups are known (see Chapter 6), the energy-level splitting can be obtained from the formula

$$r_i = \frac{1}{m} \sum_g \bar{\chi}^{(l)}(g) \chi^{(i)}(g) \tag{20.14}$$

where $\chi^{(l)}(g)$ is the character of the irreducible representation of the rotation group (or the orthogonal group) determining the symmetry of the unperturbed atom, and $\chi^{(i)}$ are the characters of the irreducible representations Γ_i of the point group in question. The quantity r_i gives the number of sub-levels whose wave functions transform by the irreducible representation Γ_i.

As an example, consider the splitting of a level with the azimuthal quantum number l in a crystal field having the symmetry of the octahedron. The characters of the representation of the group O, given by the matrices $D^{(l)}$ of the irreducible representations of O (3), will be first evaluated from (11.31). The classes C_2 and C_4 of the group O (see Chapter 6) correspond to $\varphi = \pi$ and, consequently,

$$\chi^{(l)}_{C_2} = \chi^{(l)}_{C_4} = \frac{\sin\left(l + \frac{1}{2}\right)\pi}{\sin\frac{\pi}{2}} = (-1)^l \tag{20.15}$$

The class C_3 corresponds to $\varphi = \frac{\pi}{2}$, and hence

$$\chi^{(l)}_{C_3} = \frac{\sin\left(\frac{l}{2} + \frac{1}{4}\right)\pi}{\sin\frac{\pi}{4}} = (-1)^{\left[\frac{l}{2}\right]} \tag{20.16}$$

The class C_5 corresponds to $\varphi = \frac{2\pi}{3}$, and hence

$$\chi_{C_5}^{(l)} = \frac{\sin\left(\frac{2l}{3}+\frac{1}{3}\right)\pi}{\sin\frac{\pi}{3}} = \begin{cases} 1, & l = 3m \\ 0, & l = 3m+1 \\ -1, & l = 3m+2 \end{cases} \tag{20.17}$$

Finally, for the class E we have

$$\chi_E^{(l)} = 2l+1 \tag{20.18}$$

We thus obtain the following character table for the reducible representation of the group O given by the matrices $D^{(l)}$ for $l = 0, 1, 2, 3$:

l	E	C_2	C_3	C_4	C_5
0	1	1	1	1	1
1	3	−1	1	−1	0
2	5	1	−1	1	−1
3	7	−1	−1	−1	1

If we take this together with the character table for the irreducible representations of the group O, we see that

$$\left.\begin{aligned} \mathrm{Sp}\, D^{(0)} &= \mathrm{Sp}\,\Gamma_1 \\ \mathrm{Sp}\, D^{(1)} &= \mathrm{Sp}\,\Gamma_4 \\ \mathrm{Sp}\, D^{(2)} &= \mathrm{Sp}\,\Gamma_3 + \mathrm{Sp}\,\Gamma_5 \\ \mathrm{Sp}\, D^{(3)} &= \mathrm{Sp}\,\Gamma_2 + \mathrm{Sp}\,\Gamma_4 + \mathrm{Sp}\,\Gamma_5 \end{aligned}\right\} \tag{20.19}$$

Therefore, the level-splitting scheme for the valence electron of our atom in a crystal field having the symmetry of an octahedron is as follows:

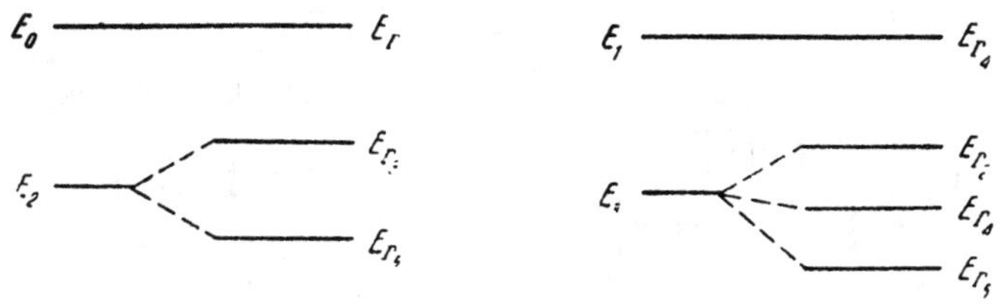

The determination of the approximate wave functions corresponding to the split energy levels is considerably facilitated by a knowledge of the linear combinations $K^{(i)}_{l\Gamma p}(\theta, \varphi)$ of spherical harmonics forming the basis of an irreducible representation of point groups (the subscript p labels the repeating irreducible representations, while the subscript i labels the basis vectors of these representations). The functions $K^{(i)}_{l\Gamma p}(\theta, \varphi)$ are called the crystal harmonics.

Table of real cubic harmonics $K^{(\varkappa)}_{lv}(\theta, \varphi)$ $\left(\int [K^{(\varkappa)}_{lv}]^2\, d\tau = 2\pi\right)$

	Γ_1	Γ_2	Γ_3
l	$K_{l\alpha}$	$K_{l\beta}$	$K_{l\gamma}$
0	$K_{0\alpha} = \frac{1}{\sqrt{2}}$		
1			
2			$K^{(1)}_{2\gamma} = P^{(0)}_2$ $K^{(2)}_{2\gamma} = \sqrt{2}\, P^{(2)}_2 \cos 2\varphi$
3		$K_{3\beta} = \sqrt{2}\, P^{(2)}_3 \sin^2 \varphi$	
4	$K_{4\alpha} = \sqrt{\frac{7}{12}}\, P^{(0)}_4 +$ $+ \sqrt{\frac{5}{6}}\, P^{(4)}_4 \cos 4\varphi$		$K^{(1)}_{4\gamma} = \sqrt{2}\, P^{(2)}_4 \cos 2\varphi$ $K^{(2)}_{4\gamma} = \sqrt{\frac{5}{12}}\, P^{(0)}_4 -$ $- \sqrt{\frac{7}{6}}\, P^{(4)}_4 \cos 4\varphi$

As an example, the table gives the cubic harmonics forming the basis of the irreducible representations of the cube group. We can use the crystal harmonics to write the wave functions for an electron level E_Γ in the form

$$\psi_i^\Gamma(\boldsymbol{r}) = \sum_{l,\,p} K_{l\Gamma p}^{(i)}(\theta,\ \varphi)\, R_{\Gamma p}(r) \tag{20.20}$$

where the functions $R_{\Gamma p}$ depend only on the length of the position vector.

Γ_4	Γ_5
$K_{l\delta}$	$K_{l\varepsilon}$
$K_{1\delta}^{(1)} = P_1^{(0)}$ $K_{1\delta}^{(2)} = \sqrt{2}\, P_1^{(1)} \cos\varphi$ $K_{1\delta}^{(3)} = \sqrt{2}\, P_1^{(1)} \sin\varphi$	
	$K_{2\varepsilon}^{(1)} = \sqrt{2}\, P_2^{(2)} \sin 2\varphi$ $K_{2\varepsilon}^{(2)} = \sqrt{2}\, P_2^{(1)} \cos\varphi$ $K_{2\varepsilon}^{(3)} = \sqrt{2}\, P_2^{(1)} \sin\varphi$
$K_{3\delta}^{(1)} = P_3^{(0)}$ $K_{3\delta}^{(2)} = \sqrt{\frac{5}{4}}\, P_3^{(3)} \cos 3\varphi -$ $- \sqrt{\frac{3}{4}}\, P_3^{(1)} \cos\varphi$ $K_{3\delta}^{(3)} = \sqrt{\frac{5}{4}}\, P_3^{(3)} \sin 3\varphi +$ $+ \sqrt{\frac{3}{4}}\, P_3^{(1)} \sin\varphi$	$K_{3\varepsilon}^{(1)} = \sqrt{2}\, P_3^{(2)} \cos 2\varphi$ $K_{3\varepsilon}^{(2)} = \sqrt{\frac{3}{4}}\, P_3^{(3)} \cos 3\varphi +$ $+ \sqrt{\frac{5}{4}}\, P_3^{(1)} \cos\varphi$ $K_{3\varepsilon}^{(3)} = \sqrt{\frac{3}{4}}\, P_3^{(3)} \sin 3\varphi -$ $- \sqrt{\frac{5}{4}}\, P_3^{(1)} \sin\varphi$
$K_{4\delta}^{(1)} = \sqrt{2}\, P_4^{(4)} \sin 4\varphi$ $K_{4\delta}^{(2)} = \sqrt{2}\sqrt{\frac{7}{8}}\, P_4^{(1)} \cos\varphi -$ $- \sqrt{2}\sqrt{\frac{1}{8}}\, P_4^{(3)} \cos 3\varphi$ $K_{4\delta}^{(3)} = \sqrt{2}\sqrt{\frac{7}{8}}\, P_4^{(1)} \sin\varphi +$ $+ \sqrt{2}\sqrt{\frac{1}{8}}\, P_4^{(3)} \sin 3\varphi$	$K_{4\varepsilon}^{(1)} = \sqrt{2}\, P_4^{(2)} \sin 2\varphi$ $K_{4\varepsilon}^{(2)} = \sqrt{2}\sqrt{\frac{1}{8}}\, P_4^{(1)} \cos\varphi +$ $+ \sqrt{2}\sqrt{\frac{7}{8}}\, P_4^{(3)} \cos 3\varphi$ $K_{4\varepsilon}^{(3)} = \sqrt{2}\sqrt{\frac{1}{8}}\, P_4^{(1)} \sin\varphi -$ $- \sqrt{2}\sqrt{\frac{7}{8}}\, P_4^{(3)} \sin 3\varphi$

Chapter 21

Selection Rules

An important application of group theory in quantum mechanics in the derivation of the selection rules. Broadly speaking, selection rules are the criteria which enable us to judge whether the matrix element of a particular operator is non-zero for a given group representation transforming this operator and the wave functions. In the theory of emission of radiation this criterion is applied to the matrix element of the operator representing the interaction with the electromagnetic field, and is used to determine the transition probabilities of a quantum-mechanical system between one stationary state and another.

21.1 General formulation of the selection rules

Consider a given set of operators $\hat{O}_\alpha$ $\alpha = 1, 2, \ldots, k$ (for example, $\hat{p}_x, \hat{p}_y, \hat{p}_z$), which transform according to the rule

$$\hat{T}_g \hat{O}_\alpha \hat{T}_g^{-1} = \sum_{\beta=1}^{k} D'_{\beta\alpha} \hat{O}_\beta \tag{21.1}$$

under a transformation g in the symmetry group G of the

quantum-mechanical system under consideration. The matrix element of the operator $\hat{O}_\alpha$ is given by

$$O_{\alpha ij} = (\psi_i, \hat{O}_\alpha \varphi_j) \tag{21.2}$$

where the functions ψ_i and φ_j form the bases of the representation $D^{(1)}$ and $D^{(2)}$ of the group G:

$$\left.\begin{aligned} \hat{T}_g \psi_i &= \sum_n D^{(1)}_{ni}(g)\,\psi_n \\ \hat{T}_g \varphi_j &= \sum_m D^{(2)}_{mj}(g)\,\varphi_m \end{aligned}\right\} \tag{21.3}$$

Since the operator $\hat{T}_g$ is unitary, we can write

$$(\psi_i, \hat{O}_\alpha \varphi_j) = (\hat{T}_g \psi_i, \hat{T}_g \hat{O}_\alpha \varphi_j) = (\hat{T}_g \psi_i, \hat{T}_g \hat{O}_\alpha \hat{T}_g^{-1} \hat{T}_g \varphi_j) \tag{21.4}$$

Substituting (21. 1) and (21. 3) into (21. 4) we obtain

$$\begin{aligned} O_{\alpha ij} &= \Big(\sum_n D^{(1)}_{ni}(g)\,\psi_n, \sum_\beta D'_{\beta\alpha}(g)\,\hat{O}_\beta \sum D^{(2)}_{mj}(g)\,\varphi_m\Big) \\ &= \sum_{n,\beta,m} \bar{D}^{(1)}_{ni}(g)\,D'_{\beta\alpha}(g)\,D^{(2)}_{mj}(g)\,O_{\beta nm} \end{aligned} \tag{21.5}$$

or

$$O_{\alpha ij} = \sum_{n,\beta,m} D_{n\beta m,\, i\alpha j} O_{\beta nm} \tag{21.6}$$

where $\| D_{n\beta m,\, i\alpha j} \|$ is the matrix of the direct product of the representations, $\bar{D}^{(1)} \times D' \times D^{(2)}$. If we sum both sides of (21. 6) over the group, and divide the result by order N of the group, we obtain

$$O_{\alpha ij} = \frac{1}{N} \sum_g \sum_{n,\beta,m} D^{(g)}_{n\beta m,\, i\alpha j} O_{\beta mn} \tag{21.7}$$

We note that the sum of matrices of any irreducible representation, other than the identity representation, over the group, is equal to the zero matrix (see Exercise 3. 7). We shall suppose that the representation $D = \bar{D}^{(1)} \times D' \times D^{(2)}$ has been decomposed into an irreducible representation as a result of the similarity transformation, i. e.

$$V D(g) V^{-1} = \sum_\lambda{}^{\oplus} D^{(\lambda)}(g) \tag{21.8}$$

where $D^{(\lambda)}(g)$ are the matrices of irreducible representations. If the right-hand side of this equation does not contain the identity representation, then

$$\sum_g \sum_\lambda^\oplus D^{(\lambda)}(g) = 0 \tag{21.9}$$

and, consequently,

$$\sum_g D(g) = V^{-1} \sum_g \sum_\lambda^\oplus D^{(\lambda)}(g) V = 0 \tag{21.10}$$

Therefore, the right-hand side of (21.7) yields zero if the representation D does not contain the identity representation.

The resulting criterion for a zero matrix element, i.e. the selection rule, can be formulated as follows: the necessary (but not sufficient) condition for the matrix element O_{aij} to be non-zero is that the direct product $\bar{D}^{(1)} \times D' \times D^{(2)}$ should contain the identity representation.

If, in particular, the operator $\hat{O}$ is the unit operator and, consequently, transforms by the identity representation, then the above selection rule will express the orthogonality of the functions transforming the inequivalent irreducible representations (see Section 5.3).

21.2 Selection rules for the absorption and emission of radiation

To illustrate the application of the theory derived in the previous section, let us establish the selection rules for the emission and absorption of light by atoms. We shall restrict our attention to the dipole approximation in which the transition probability between states A and B is proportional to the square of the modulus of the matrix element

$$\boldsymbol{r}_{AB} = \int \bar{\Psi}_A(x) \sum_j \boldsymbol{r}_j \Psi_B(x)\, d\tau \tag{21.11}$$

where $\boldsymbol{r}_j$ are the position vectors of the atomic electrons, x represents the set of space and spin coordinates and the integral with respect to $d\tau$ represents integration with respect to the space variables and summation over spin variables.

Consider the case of $L-S$ coupling. We shall suppose that the wave functions Ψ_A and Ψ_B for the initial and final states have been constructed from one-electron wave functions corresponding to the given configuration. The matrix element (21. 11) can then be expressed in terms of the one-electron matrix elements and

$$\int \bar{\psi}_l(\mathbf{r})\, \mathbf{r} \psi_{l'}(\mathbf{r})\, d\mathbf{r} \tag{21.12}$$

where ψ_l is the one-electron wave function with azimuthal quantum number l. Let us establish the selection rules due to the symmetry of the system under inversion. The position vector $\mathbf{r}$ transforms in accordance with an odd representation of the inversion group. The wave function ψ_l transforms by an even representation if l is even and an odd representation if l is odd. To ensure that the integral (21. 12) is non-zero it is necessary that the integrand should transform by an even (identity) representation, i.e. l and l' should have different parity.

Consider now the selection rules for the matrix element (21. 12) due to symmetry under rotation. It is clear that the integrand in (21. 12) transforms in accordance with the representation $D^{(l)} \times D^{(1)} \times D^{(l')}$ under rotations. From the Clebsch-Gordan formula we may conclude that this representation contains the identity representation only if

$$l = l' \quad \text{or} \quad l = l' \pm 1$$

If we recall the parity selection rules obtained earlier, we see that the matrix element (21. 12) can be non-zero only if

$$\Delta l = l' - l = \pm 1 \tag{21.13}$$

Since the dipole-moment operator in (21. 11) is a sum of one-electron operators, the integration yields a non-zero result only if the initial and final configurations differ by only the azimuthal quantum number.

Let us now find the selection rules for the matrix element (21. 11) without the use of the one-electron approximation.

Let the states A and B be characterized by quantum numbers L, S, J and L', S', J', respectively. By considering the selection rules for (21. 11) due to the rotation group, we obtain

$$\Delta L = L' - L = 0, \pm 1 \tag{21.14}$$

$$\Delta J = J' - J = 0, \pm 1 \tag{21.15}$$

We note, however, that the transitions $J=0 \rightarrow J'=0$ and $L=0 \rightarrow L'=0$ are forbidden. This is obvious if the representation $D^{(0)} \times D^{(1)} \times D^{(0)}$, which transforms the product of the three factors in (21. 11), is decomposed into irreducible parts in accordance with the Clebsch-Gordan rule. Next, the dipole-moment operator $\sum \boldsymbol{r}_i$ is symmetric under the interchange of the position vectors $\boldsymbol{r}_i$. Therefore, to ensure that the representation transforming the integrand in (21. 12) should contain the identity representation, it is necessary for the functions Ψ_A and Ψ_B to transform in accordance with inequivalent representations of the permutation group. However, since the representation of the permutation group for the space variables $\boldsymbol{r}_i$ is uniquely related to the total spin eigenvalues (see Chapter 17), we find that

$$\Delta S = S' - S = 0 \tag{21.16}$$

21.3 Selection rules for Raman scattering by molecules

As a further example consider the selection rules for Raman scattering by molecules. We shall restrict our attention to a model of a molecule with two non-degenerate electron terms, each of which has its own vibrational structure. The vibrational sub-levels for a normal vibration of the molecule are shown in Fig. 17. Raman scattering is a second-order

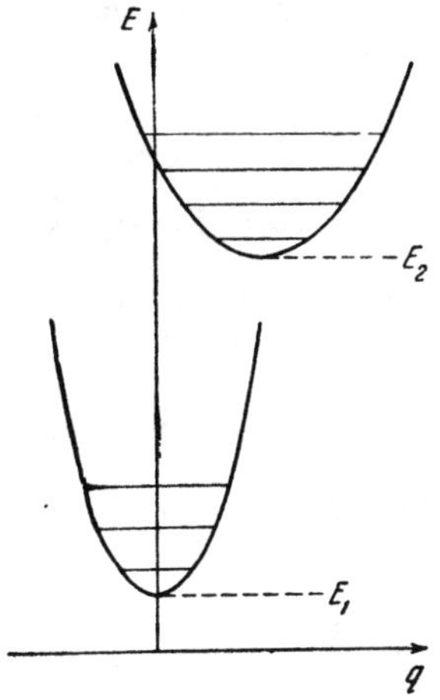

Fig. 17

process as far as interaction with the electromagnetic field is concerned, and involves virtual transitions to the vibrational sub-levels of an excited state.

Suppose that the molecule absorbs a photon of frequency ω_0 and polarization α and emits a photon of frequency ω_1 and polarization β. We shall assume that the molecule changes only its vibrational state and continues to reside in the ground electronic state. The probability of this process is proportional to the square of the modulus of the following quantity:

$$K = \sum \frac{\langle 1n | M_\alpha | 2m \rangle \langle 2m | M_\beta | 1n' \rangle}{E_2 + \varepsilon_m - E_1 - \varepsilon_n - \hbar\omega_0}$$

$$+ \sum_m \frac{\langle 1n | M_\alpha | 2m \rangle \langle 2m | M_\beta | 1n' \rangle}{E_2 + \varepsilon_m - E_1 - \varepsilon_n + \hbar\omega_0} \qquad (21.17)$$

where $\langle 1n | M_\alpha | 2m \rangle$ represents the matrix element of the electronic dipole-moment operator of the molecule, evaluated using wave functions of the n-th vibrational sub-level of the electronic ground state and the m-th vibrational sub-level of the electronic excited state; the quantities

E_2 and E_1 represent the minima of the excited and ground terms, respectively, and ε_m and ε_n are the energies of the vibrational sub-levels. The frequency ω_1 of the emitted photon is determined by the conservation of energy

$$E_1 + \varepsilon_n + \hbar\omega_0 = E_2 + \varepsilon_{n'} + \hbar\omega_1 \qquad (21.18)$$

We shall consider non-resonant Raman scattering, i.e. we shall assume that

$$E_2 - E_1 - \hbar\omega_0 \gg \varepsilon_m - \varepsilon_n \qquad (21.19)$$

If we neglect the small quantity $\varepsilon_m - \varepsilon_n$ in the denominators of (21.17), we find that

$$K = \left\{ \frac{1}{E_2 - E_1 - \hbar\omega_0} + \frac{1}{E_2 - E_1 + \hbar\omega_0} \right\} \times \sum_m \langle 1n | M_\alpha | 2m \rangle \langle 2m | M_\beta | 1n' \rangle \qquad (21.20)$$

In the adiabatic approximation, the wave function of a molecule can be written in the form

$$\Psi_{im}(\boldsymbol{r}, \boldsymbol{q}) = \psi_i(\boldsymbol{r}, \boldsymbol{q}) \Phi_m^{(i)}(\boldsymbol{q}) \qquad (21.21)$$

where $\psi_i(\boldsymbol{r}, \boldsymbol{q})$ is the electron wave function, i.e. the solution of the Schroedinger equation for a fixed configuration of the nuclei, which depends on the displacements $\boldsymbol{q}$, and $\Phi_m^{(i)}(\boldsymbol{q})$ is the vibrational wave function. We can now write

$$K = \frac{2(E_2 - E_1)}{(E_2 - E_1)^2 - (\hbar\omega_0)^2} \sum_m \langle n | M_{\alpha 12}(\boldsymbol{q}) | m \rangle \langle m | M_{\beta 21}(\boldsymbol{q}) | n' \rangle$$

$$= \frac{2(E_2 - E_1)}{(E_2 - E_1)^2 - (\hbar\omega_0)^2} \sum \langle n | M_{\alpha 12}(\boldsymbol{q}) M_{\beta 21}(\boldsymbol{q}) | n' \rangle \qquad (21.22)$$

where

$$M_{\alpha ij}(\boldsymbol{q}) = \int \bar{\psi}_i(\boldsymbol{r}, \boldsymbol{q}) r_\alpha \psi_j(\boldsymbol{r}, \boldsymbol{q})\, d\boldsymbol{r} \qquad (21.23)$$

To find the selection rules for (21.22), let us first determine the transformation rule for the electronic matrix elements $M_{\alpha ij}(\boldsymbol{q})$ under transformations of the displacements $\boldsymbol{q}$ giving a representation D of the symmetry group G of the molecule in the electronic ground state. If g is an element of G, then

$$\psi_j(g^{-1}\boldsymbol{r},\ D(g^{-1})\boldsymbol{q}) = e^{i\delta_j(g)}\psi_j(\boldsymbol{r},\ \boldsymbol{q}) \tag{21.24}$$

and, in particular, if $\boldsymbol{q} = \boldsymbol{0}$, we have

$$\psi_j(g^{-1}\boldsymbol{r},\ \boldsymbol{0}) = e^{i\delta_j(g)}\psi_j(\boldsymbol{r},\ \boldsymbol{0}) \tag{21.25}$$

Hence, it follows that the factors $e^{i\delta_j(g)}$ form a representation of the group G which transforms the electronic wave functions for the equilibrium configuration of the nuclei. We are interested in the quantity $M_{\alpha 12}(D(g^{-1})\boldsymbol{q})$, whose explicit form is

$$M_{\alpha 12}(D(g^{-1})\boldsymbol{q}) = \int \bar{\psi}_1(\boldsymbol{r},\ D(g^{-1})\boldsymbol{q})\, \boldsymbol{r}_\alpha \psi_2(\boldsymbol{r},\ D(g^{-1})\boldsymbol{q})\, d\boldsymbol{r} \tag{21.26}$$

If we change the integration variable so that

$$\boldsymbol{r} = g^{-1}\boldsymbol{r}' \tag{21.27}$$

we obtain

$$M_{\alpha 12}(D(g^{-1})\boldsymbol{q})$$
$$= \int \bar{\psi}_1(g^{-1}\boldsymbol{r}',\ D(g^{-1})\boldsymbol{q}) \sum_{\gamma} g^{-1}_{\alpha\gamma} \boldsymbol{r}'_\gamma \psi_2(g^{-1}\boldsymbol{r}',\ D(g^{-1})\boldsymbol{q})\, d\boldsymbol{r} \tag{21.28}$$

Using (21.24) and the orthogonality of the matrix $\|g_{\alpha\beta}\|$, we find that

$$M_{\alpha 12}(D(g^{-1})\boldsymbol{q}) = e^{-i\delta_1(g)} e^{i\delta_2(g)} \sum_{\gamma} g_{\gamma\alpha} \int \bar{\psi}_1(\boldsymbol{r}',\ \boldsymbol{q})\, \boldsymbol{r}'_\gamma \psi_2(\boldsymbol{r}',\ \boldsymbol{q})\, d\boldsymbol{r}'$$

$$= e^{-i\delta_1(g)} e^{i\delta_2(g)} \sum_{\gamma} g_{\gamma\beta} M_{\gamma 12}(\boldsymbol{q}) \tag{21.29}$$

It is now a simple matter to find the transformation rule for the operator $O_{\alpha\beta}(\boldsymbol{q}) = M_{\alpha 12} M_{\beta 21}$ in (21.22). We have

$$O_{\alpha\beta}(D(g^{-1})\boldsymbol{q}) = M_{\alpha 12}(D(g^{-1})\boldsymbol{q})\, M_{\beta 21}(D(g^{-1})\boldsymbol{q})$$
$$= \sum_{\gamma,\ \gamma'} g_{\gamma\alpha} g_{\gamma'\beta} M_{\alpha 12}(\boldsymbol{q})\, M_{\beta 21}(\boldsymbol{q}) = \sum_{\gamma,\ \gamma'} g_{\gamma\alpha} g_{\gamma'\beta} O_{\alpha\beta} \tag{21.30}$$

It is clear that the operator $O_{\alpha\beta}(\boldsymbol{q})$ transforms in accordance with the tensor representation of G. If the matrix elements $M_{\alpha 12}$ of the dipole moment are real, then we have the additional condition

$$O_{\alpha\beta} = O_{\beta\alpha} \tag{21.31}$$

In this case, the six independent quantities will transform in accordance with the symmetric square $[g^2]$ of the vector representation. We can now use this result to formulate the selection rule for Raman scattering: a transition is allowed if the direct product $D^{(n)} \times D^{(n')} \times [g^2]$, where $D^{(n)}$ and $D^{(n')}$ are the representations transforming the oscillator wave functions, contains the identity representation.

21.4 Matrix elements constructed in terms of the functions of a given basis

If the wave functions used to construct the matrix element of a Hermitian operator belong to the basis of a single representation, it is possible to establish more stringent selection rules as compared with those derived above. These new selection rules are important when it is necessary to calculate the matrix elements with wave functions belonging to a given energy level. Consider, to begin with, the case where the wave functions are real and spin-independent. The matrix element of the Hermitian operator $\hat{O}_\alpha$ in which we are interested can be written in the form

$$\langle i | \hat{O}_\alpha | j \rangle = \int \psi_i \hat{O}_\alpha \psi_j \, d\tau \tag{21.32}$$

Suppose the functions ψ_k transform in accordance with the representation D, while the operator $\hat{O}_\alpha$ transforms in accordance with the representation D' of a group G. To ensure that the matrix element (21. 32) is non-zero, it is then necessary that the direct product $D \times D \times D'$ should contain the identity representation. It is, however, possible to obtain certain additional limitations if we take into account the symmetry of the matrix element under the interchange of the symbols i and j. We emphasize that when the functions on the left and on the right of the operator belong to the same basis of a particular representation, the interchange of the subscripts yields again the same set of matrix elements, and this does not occur if the functions belong to different bases. We can write

$$\langle i|\hat{O}_\alpha|j\rangle = \int \psi_i\hat{O}_\alpha\psi_j\,d\tau = \int (\bar{\hat{O}}_\alpha\psi_i)\,\psi_j\,d\tau$$

$$= \pm\int \psi_j\hat{O}_\alpha\psi_i\,d\tau = \pm\langle j|\hat{O}_\alpha|i\rangle \qquad (21.33)$$

where the positive and negative signs refer, respectively, to a real or purely imaginary operator $\hat{O}_\alpha$ (for example, the operator representing the interaction between the electron and an electric field is real, while the operator for the interaction with the magnetic field is purely imaginary). From the symmetry or antisymmetry of matrix elements we can conclude that the representation under which they transform must contain the symmetric or antisymmetric square of the representation D as a factor. Using the notation of Chapter 16, we can write the representation transforming the matrix elements in the form

$$\left.\begin{array}{ll} [D^2]\times D' & \text{for a real operator} \\ \{D^2\}\times D' & \text{for an imaginary operator} \end{array}\right\} \qquad (21.34)$$

The selection rules can now be formulated as follows: to ensure that the matrix element (21.32) is non-zero, it is necessary for the representation (21.34) to contain the identity representation.

The selection rules must be modified if we take the number of states into account. We shall restrict our attention to the case where the energy operator for the given system is invariant under time inversion, i.e. it does not contain interactions with the magnetic field (see Chapter 13). Let ψ be an eigenfunction of the corresponding Hamiltonian so that $\hat{\Theta}\psi$ is also an eigenfunction of this Hamiltonian with the same eigenvalue, where $\hat{\Theta}$ is the time reversal operator. We recall that for an n-electron system

$$\hat{\Theta} = \hat{S}_{1y}\hat{S}_{2y}\ldots\hat{S}_{ny}K \qquad (21.35)$$

where $\hat{S}_{iy}$ is the spin-projection operator for the i-th electron along the y-axis, and K represents complex conjugation.

The operator $\hat{\Theta}$ can readily be shown to commute with the infinitesimal rotation operators and, consequently, with any rotation. It follows, however, that if the function ψ transforms by a representation D of the rotation group or its sub-group, then the function $\hat{\Theta}\psi$ will transform by the complex conjugate representation $\bar{D}$. In fact,

$$\hat{T}_g\hat{\Theta}\psi_i = \hat{\Theta}\hat{T}_g\psi_i = \hat{\Theta}\sum_j D_{ji}(g)\psi_j = \sum \bar{D}_{ji}(g)\hat{\Theta}\psi_j \qquad (21.36)$$

Hence, it follows that the eigenfunctions of the Hamiltonian corresponding to a single eigenvalue transform either by a real representation or by a representation which is equivalent to its complex conjugate. In the latter case, it can be written in the form $D^{(s)} \oplus \bar{D}^{(s)}$, where $D^{(s)}$ is a representation of the group under consideration. The similarity transformation which makes the representation $D^{(s)} \oplus \bar{D}^{(s)}$ real can readily be found.

Consider the transformation rule for the wave functions when the operator $\hat{\Theta}$ acts upon them. We have

$$\hat{\Theta}\psi_i = \sum_j \theta_{ji}\psi_j \qquad (21.37)$$

where $\|\theta_{ji}\|$ is a unitary matrix: $\Theta^+ = \Theta^{-1}$. We know (see Chapter 13) that

$$\hat{\Theta}^2 = \varkappa \qquad (21.38)$$

where $\varkappa = 1$ for an even number of electrons and $\varkappa = -1$ for an odd number. The corresponding matrix equation is

$$\Theta\bar{\Theta} = \varkappa E \qquad (21.39)$$

and hence

$$\Theta^{-1} = \varkappa\bar{\Theta} \qquad (21.40)$$

Since Θ is unitary, we have

$$\Theta^{-1} = \Theta^+ = \varkappa\bar{\Theta} \qquad (21.41)$$

or

$$\Theta^* = \varkappa\Theta \qquad (21.42)$$

where Θ^* is the transposed matrix. This property of the matrix Θ will be used to determine the symmetry of the matrix element $O_{\alpha ij} = (\psi_i, \hat{O}_\alpha\psi_j)$ under the interchange of the symbols i and j.

We can now write

$$O_{\alpha ij} = (\hat{\Theta}\psi_i, \hat{\Theta}\hat{O}_\alpha\psi_j) = (\hat{\Theta}\psi_i, \hat{\Theta}\hat{O}_\alpha\hat{\Theta}^{-1}\hat{\Theta}\psi_j)$$

$$= \pm \sum_{m, n} \bar{O}_{\alpha mn}\theta_{mi}\bar{\theta}_{nj} \qquad (21.43)$$

where the positive sign corresponds to $\hat{\Theta}\hat{O}_\alpha\hat{\Theta}^{-1} = \hat{O}_\alpha$ and the negative sign to $\hat{\Theta}\hat{O}_\alpha\hat{\Theta}^{-1} = -\hat{O}_\alpha$. Since the operator $\hat{O}_\alpha$ is Hermitian $\bar{O}_{\alpha mn} = O_{\alpha nm}$ and the matrix Θ is unitary $\Theta^{-1} = \Theta^+$, we can rewrite (21.43) in the form

$$\sum_i \bar{\theta}_{mi} O_{\alpha ij} = \pm \sum_n \bar{\theta}_{nj} O_{\alpha nm} \qquad (21.44)$$

or, in view of (21.42),

$$\sum_i \bar{\theta}_{mi} O_{\alpha ij} = \pm \varkappa \sum_n \bar{\theta}_{jn} O_{\alpha nm} \qquad (21.45)$$

If instead of the matrix elements $O_{\alpha ij}$ we introduce the quantities

$$\tilde{O}_{\alpha mj} = \sum_i \bar{\theta}_{mi} O_{\alpha ij} \qquad (21.46)$$

we have from (21.45)

$$\tilde{O}_{\alpha mj} = \pm \varkappa \tilde{O}_{\alpha jm} \qquad (21.47)$$

Let us find the selection rules for the quantities $\tilde{O}_{\alpha mj}$. We have shown that the representation transforming the wave function ψ_i can be regarded as real and, therefore, the quantities $\tilde{O}_{\alpha mj}$ transform as the components of a tensor. In view of (21.47) we find that for an even number of electrons ($\varkappa = 1$) and for an operator $\hat{O}_\alpha$ commuting with $\hat{\Theta}$, or for an odd number of electrons ($\varkappa = -1$) and $\hat{O}_\alpha$ anticommuting with $\hat{\Theta}$, the quantities $\tilde{O}_{\alpha mj}$ are the components of a symmetric tensor. For an odd number of electrons and $\hat{O}_\alpha$ commuting with $\hat{\Theta}$, or for an even number of electrons and $\hat{O}_\alpha$ anticommuting with $\hat{\Theta}$, the quantities $\tilde{O}_{\alpha mj}$ are the components of an antisymmetric tensor. Consequently, the representation transforming the quantities $\tilde{O}_{\alpha mj}$ in the former case are of the form

$$[D^2] \times D' \tag{21.48}$$

while in the latter case they are of the form

$$\{D^2\} \times D' \tag{21.49}$$

We can now conclude that to ensure that the quantities $\tilde{O}_{\alpha ij}$ are non-zero, it is necessary that the representations (21.48) and (21.49) should contain the identity representation. These selection rules remain valid for matrix elements $O_{\alpha ij}$ which are related to $\tilde{O}_{\alpha mj}$ by a unitary transformation.

As an example of the use of selection rules derived here, let us consider the stability of symmetric configurations in molecules.

21.5 The Jahn-Teller theorem

Consider the Schroedinger equation for the electronic state of a molecule:

$$\left[-\frac{\hbar^2}{2m}\Delta_r + V(\boldsymbol{r}, \boldsymbol{R})\right]\psi(\boldsymbol{r}, \boldsymbol{R}) = E(\boldsymbol{R})\psi(\boldsymbol{r}, \boldsymbol{R}) \tag{21.50}$$

where $\boldsymbol{R}$ represents the positions of the nuclei and $\boldsymbol{r}$ those of the electrons. The eigenvalue of the electron energy $E(\boldsymbol{R})$ is a function of the coordinates of the nuclei, and the minimum of this eigenvalue corresponds to the equilibrium configuration of the nuclei for the given electronic state. Let us suppose that we do not know the exact equilibrium positions of the nuclei. We shall take the symmetric configuration $\boldsymbol{R}^{(1)}$ as the zero-order approximation, and will suppose that it is close to the equilibrium configuration $\boldsymbol{R}^{(0)}$, i.e. $\boldsymbol{R}_i^{(0)} = \boldsymbol{R}_i^{(1)} + \Delta \boldsymbol{R}_i$, where $\Delta \boldsymbol{R}_i$ is a small displacement of the i-th nucleus. The equilibrium configuration $\boldsymbol{R}^{(0)}$ will be sought using the perturbation theory. The Schroedinger equation (21.50) for the configuration $\boldsymbol{R}^{(0)}$ can be written in the form

$$\left[-\frac{\hbar^2}{2m}\Delta_r + V(\boldsymbol{r}, \boldsymbol{R}^{(1)}) + W(\boldsymbol{r})\right]\psi(\boldsymbol{r}) = E\psi(\boldsymbol{r}) \tag{21.51}$$

where

$$W(\boldsymbol{r}) = V(\boldsymbol{r}, \boldsymbol{R}^{(0)}) - V(\boldsymbol{r}, \boldsymbol{R}^{(1)}) \cong \sum_i \left(\frac{\partial V(\boldsymbol{r}, \boldsymbol{R})}{\partial \boldsymbol{R}_i}\right)_{\boldsymbol{R}^{(1)}} \Delta \boldsymbol{R}_i \tag{21.52}$$

can be regarded as a perturbation.

We shall denote the Cartesian components of the displacements of the nuclei relative to the configuration $\boldsymbol{R}^{(1)}$ by x_β, where β labels the Cartesian axes and the nuclei. We know (see Chapter 5) that the displacements x_β transform in accordance with a representation of the symmetry group G of the configuration $\boldsymbol{R}^{(1)}$:

$$x'_\alpha = \sum_\beta D_{\alpha\beta}(g)\, x_\beta \tag{21.53}$$

Let us now transform from the displacements x_β to the symmetrized displacements q_β which transform in accordance with an irreducible representation of G:

$$q_\beta = \sum_\alpha b_{\alpha\beta} x_\alpha \tag{21.54}$$

The perturbation W can now be expressed in terms of the coordinates q_α:

$$W = \sum_\alpha \left(\frac{\partial V}{\partial q_\alpha}\right)_{q=0} q_\alpha \tag{21.55}$$

We can now find a correction to the energy eigenvalue $E(\boldsymbol{R}^{(1)})$ due to the perturbation W. Let the eigenvalue $E(\boldsymbol{R}^{(1)})$ be s-fold degenerate, and suppose that the corresponding eigenfunctions transform by an irreducible representation D of the symmetry group G of the configuration $\boldsymbol{R}^{(1)}$. We shall denote these functions by $\psi_1, \psi_2, \ldots, \psi_s$, and will suppose that they are orthonormal. We note that the first-order perturbation-theory correction to the energy is determined by the roots of the secular equation

$$|w_{ik} - \lambda\delta_{ik}| = 0 \qquad (21.56)$$

In our case,

$$w_{ik} = \int \overline{\psi}_i(r) \sum_\alpha \left(\frac{\partial V}{\partial q_\alpha}\right)_0 q_\alpha \psi_k(r)\, dr \qquad (21.57)$$

We see that the roots of (21.56) will be functions of q_α, the values of which for the equilibrium configurations are chosen by the condition of minimum $E(q)$. By inverting (21.54) with the values of q_α found in this way, we obtain the displacements of the nuclei, and hence the possible equilibrium configurations.

In Chapter 6 we found the coordinates q_α for a molecule with cubic symmetry, and gave their geometrical interpretation. In particular, we saw that only the coordinate q_1 which transformed by the identity representation of a point group corresponded to a change in the position of the nuclei which did not modify the symmetry of the molecule.

Let us now use group-theoretical considerations to find the condition that the matrix elements

$$\int \overline{\psi}_i(r) \left(\frac{\partial V(r)}{\partial q_\alpha}\right)_0 \psi_k(r)\, dr \qquad (21.58)$$

are zero. To do this, we must first establish the transformation rule for the functions $\left(\frac{\partial V(r)}{\partial q_\alpha}\right)_0 \equiv W_\alpha(r)$ when a transformation from the point group G is applied to the electron coordinates. Consider the set of derivatives $\left(\frac{\partial V}{\partial q_\alpha}\right)_0$, where the q_α transform by an irreducible representation Γ of G. We shall show that these derivatives transform by the same representation Γ. We know that the perturbation W is invariant under any simultaneous orthogonal transformation of the positions of the nuclei and electrons. We therefore

have

$$W=\sum_{\alpha} W_{\alpha}(\boldsymbol{r})\, q_{\alpha}=\sum_{\beta} W_{\beta}(\boldsymbol{r}')\, q'_{\beta} \tag{21.59}$$

where

$$\boldsymbol{r}'=g\boldsymbol{r}, \qquad q'_{\beta}=\sum_{\alpha}\Gamma_{\beta\alpha}(g)\, q_{\alpha} \tag{21.60}$$

We thus find that

$$\sum_{\beta} W_{\beta}(\boldsymbol{r}')\, q'_{\beta}=\sum_{\beta,\alpha} W_{\beta}(\boldsymbol{r}')\,\Gamma_{\beta\alpha} q_{\alpha}=\sum_{\alpha} W_{\alpha}(\boldsymbol{r})\, q_{\alpha} \tag{21.61}$$

and hence

$$W_{\alpha}(\boldsymbol{r})=\sum_{\beta}\Gamma_{\beta\alpha} W_{\beta}(g\boldsymbol{r}) \tag{21.62}$$

Substituting $\boldsymbol{r}\to g^{-1}\boldsymbol{r}$, we finally obtain

$$W_{\alpha}(g^{-1}\boldsymbol{r})=\sum_{\beta}\Gamma_{\beta\alpha}(g)\, W_{\beta}(\boldsymbol{r}) \tag{21.63}$$

which was to be proved.

Since the Hamiltonian for our problem does not contain the spin operators, we can use the selection rules formulated in the previous section. The perturbation operator W_{α} is real. We can therefore expect that the matrix element (21.58) is non-zero if the representation $[D^2]$ contains Γ. If the electronic state is not degenerate, i.e. D is one-dimensional, then $[D^2]$ is the identity representation, and the only displacements which are possible are the fully symmetric displacements which do not change the symmetry of the molecule. If the electronic state is degenerate, however, then detailed studies of all the possible types of symmetry of the molecules show that it is always possible to find a displacement which is not fully symmetric and transforms by a representation contained in $[D^2]$ (H. A. Jahn and E. Teller, Proc. R. Soc., A, V 161, p.220 (1937)). Linear molecules are the only exception. It is clear that it is always possible to choose the value of this displacement so that the corresponding correction to the energy is negligible. In fact, if for a particular value $q=\Delta$ the correction is positive, then for $q=-\Delta$ it should be negative.

Hence, it follows that if for a particular symmetric configuration of the nuclei the electronic state is degenerate, the molecule 'tends to reduce its symmetry so as to remove the degeneracy' (this is the Jahn-Teller theorem).

Exercises

21.1. Find the selection rules for the emission and absorption of light by atoms in the $j - j$ coupling approximation.

21.2. Show that the tetrahedral configuration of the CH_4 molecule is unstable for degenerate electronic states.

Chapter 22

The Lorentz Group and its Irreducible Representations

22.1 The general Lorentz group

According to the theory of relativity, the space coordinates and the time in two reference frames in uniform relative motion are related by a linear transformation which we shall call a proper Lorentz transformation. If the time and the space coordinates in one frame are denoted by x_0, x_1, x_2, x_3 and in the other by y_0, y_1, y_2, y_3, then

$$y_\alpha = \sum_{\beta=0}^{3} \Lambda_{\alpha\beta} x_\beta \tag{22.1}$$

where $\|\Lambda_{\alpha\beta}\|$ is the Lorentz transformation matrix. To establish some of the properties of this matrix, we shall use the fact that the quadratic form

$$\varphi = -x_0^2 + x_1^2 + x_2^2 + x_3^2 \tag{22.2}$$

is invariant under the Lorentz transformation (we have assumed the velocity of light to be equal to unity). If the four-dimensional vector with the components x_0, x_1, x_2, x_3 is written in the form of the column matrix

$$X = \begin{pmatrix} x_0 \\ x_1 \\ x_2 \\ x_3 \end{pmatrix}$$

then the quadratic form (22. 2) can be written as

$$\varphi = X^*FX \tag{22.3}$$

where the asterisk represents the transposed matrix and

$$F = \begin{Vmatrix} -1 & 0 & 0 & 0 \\ 0 & 1 & 0 & 0 \\ 0 & 0 & 1 & 0 \\ 0 & 0 & 0 & 1 \end{Vmatrix} \tag{22.4}$$

The invariance of the quadratic form given by (22. 2) under the Lorentz transformation Λ can now be written in the form

$$X^*FX = Y^*FY \tag{22.5}$$

where

$$Y = \Lambda X \tag{22.6}$$

Substituting (22. 6) in (22. 5), we obtain

$$(\Lambda X)^* F\Lambda X = X^*\Lambda^*F\Lambda X = X^*FX \tag{22.7}$$

Consequently,

$$\Lambda^*F\Lambda = F \tag{22.8}$$

We thus see that the condition given by (22. 8) is satisfied by a class of real linear transformations which is broader than the proper Lorentz transformations. All such transformations will be called general Lorentz transformations. It is readily verified that the general Lorentz transformations form a group – the general Lorentz group L.

Let us calculate the determinant of the right- and left-hand sides of (22. 8). We have

$$(\det \Lambda)^2 = 1 \tag{22.9}$$

and hence

$$\det \Lambda = \pm 1 \tag{22.10}$$

Thus, the general Lorentz group can be decomposed into two parts: L_+ a set of matrices with the determinant equal to unity, and L_- a set of matrices with the determinant equal to -1. It is clear that L_+ itself forms a group, whereas L_- does not.

The elements at the intersection of the first row and the first column of the matrix in (22.8) are equal, i.e.

$$\Lambda_{00}^2 - \Lambda_{10}^2 - \Lambda_{20}^2 - \Lambda_{30}^2 = 1 \tag{22.11}$$

Similarly, from the equation

$$\Lambda F \Lambda^* = F \tag{22.12}$$

which can be deduced from (22.8) with the aid of transposition of matrices, we find that

$$\Lambda_{00}^2 - \Lambda_{01}^2 - \Lambda_{02}^2 - \Lambda_{03}^2 = 1 \tag{22.13}$$

From (22.11) or (22.13) it follows that

$$\Lambda_{00}^2 \geqslant 1 \tag{22.14}$$

and hence either $\Lambda_{00} \geqslant 1$ or $\Lambda_{00} \leqslant -1$. According to this criterion, each of the sets L_+ and L_- can in its turn be split into two sets. In this way, the general Lorentz group splits into the four sets

L_+^+	L_+^-	L_-^+	L_-^-
$\Lambda_{00} \geqslant 1$	$\Lambda_{00} \leqslant -1$	$\Lambda_{00} \geqslant 1$	$\Lambda_{00} \leqslant -1$
$\det \Lambda = 1$	$\det \Lambda = 1$	$\det \Lambda = -1$	$\det \Lambda = -1$

It is clear that these sets cannot be related to each other through a continuous variation of the elements of the Lorentz matrix. The set L_+^+ in itself forms a group. The proof of this is based on the following property: matrices for which $\Lambda_{00} \geqslant 1$ form a group. In fact, let Λ and $\tilde{\Lambda}$ be two such matrices. The element of the matrix $\Lambda\tilde{\Lambda}$ in which we are interested is given by

$$\{\Lambda\tilde{\Lambda}\}_{00} = \Lambda_{00}\tilde{\Lambda}_{00} + \Lambda_{01}\tilde{\Lambda}_{10} + \Lambda_{02}\tilde{\Lambda}_{20} + \Lambda_{03}\tilde{\Lambda}_{30} \tag{22.15}$$

Using the Schwartz inequality and Equations (22.11) and

(22. 13), we have

$$(\Lambda_{01}\tilde{\Lambda}_{10}+\Lambda_{02}\tilde{\Lambda}_{20}+\Lambda_{03}\tilde{\Lambda}_{30})$$

$$\leqslant(\Lambda_{01}^2+\Lambda_{02}^2+\Lambda_{03}^2)(\tilde{\Lambda}_{10}^2+\tilde{\Lambda}_{20}^2+\tilde{\Lambda}_{30}^2)<\Lambda_{00}^2\tilde{\Lambda}_{00}^2 \qquad (22.16)$$

Hence, it follows that $\{\Lambda\tilde{\Lambda}\}_{00}>0$. Since the matrix $\Lambda\tilde{\Lambda}$ belongs to the Lorentz group, we finally find that

$$\{\Lambda\tilde{\Lambda}\}_{00}\geqslant 1 \qquad (22.17)$$

which was to be proved.

It can be shown that any two matrices belonging to a given set can be transformed one into the other in a continuous fashion. Hence, it follows, in particular, that the group $L_+^\dagger$ which contains the unit matrix will also contain all the proper Lorentz transformations. The group $L_+^\dagger$ is called the proper Lorentz group. If we consider the three general Lorentz matrices

$$F=\begin{pmatrix}-1&0&0&0\\0&1&0&0\\0&0&1&0\\0&0&0&1\end{pmatrix} \qquad S=\begin{pmatrix}1&0&0&0\\0&-1&0&0\\0&0&-1&0\\0&0&0&-1\end{pmatrix}$$

$$I=\begin{pmatrix}-1&0&0&0\\0&-1&0&0\\0&0&-1&0\\0&0&0&-1\end{pmatrix}$$

we can decompose the general Lorentz group into the following cosets:

$$L_+^\dagger,\quad FL_+^\dagger,\quad SL_+^\dagger,\quad IL_+^\dagger$$

It is clear that the last three cosets coincide with $L_-^\downarrow$, $L_-^\dagger$, $L_+^\downarrow$.

22.2 Relation between the Lorentz group and the four-dimensional rotation group

We shall now show that the Lorentz group is uniquely related to the four-dimensional rotation group $O^+(4)$ in the neighborhood of the unit element. This will help us later to deduce all the finite-dimensional irreducible representations of the proper Lorentz group.

Let us first find the number of parameters or the number of independent elements of the Lorentz transformation matrix. We have seen that the Lorentz matrices must satisfy (22.8). Since the matrix $\Lambda^* F \Lambda$ is symmetric, i.e. $(\Lambda^* F \Lambda)^* = \Lambda^* F \Lambda$, it follows that (22.8) is equivalent to 10 conditions imposed on the matrix elements. Hence, it follows that of the 16 elements of the Lorentz matrix only six are independent. The Lorentz transformation is, therefore, a six-parameter transformation. These parameters can be the three components of the relative velocity and the three Euler angles defining the mutual orientation of the coordinate systems. It will be more convenient, however, to take the independent parameters to be the angles of rotation in the planes $(x_0 x_1)$, $(x_0 x_2)$, $(x_0 x_3)$, $(x_1 x_2)$, $(x_1 x_3)$, $(x_2 x_3)$. These parameters will be denoted by φ_{01}, φ_{02}, φ_{03}; ψ_{12}, ψ_{13}, ψ_{23}, while the corresponding infinitesimal matrices will be denoted by B_{01}, B_{02}, B_{03}; B_{12}, B_{13}, B_{23}.

Consider the four-dimensional rotation group $O^+(4)$. The coordinates of the space in which the transformations in $O^+(4)$ operate will also be denoted by x_0, x_1, x_2, x_3. In contrast to (22.1), the positive-definite quadratic form

$$x_0^2 + x_1^2 + x_2^2 + x_3^2 \tag{22.18}$$

will now be invariant for this group.

Consider an arbitrary transformation in the group of four-dimensional rotations which preserves the coordinates x_2 and x_3. It is clear that this is an ordinary rotation in the plane $(x_0 x_1)$, which can be written in the form

$$x_0' = x_0 \cos\varphi - x_1 \sin\varphi, \quad x_1' = x_0 \sin\varphi + x_1 \cos\varphi \tag{22.19}$$

Let us consider these equations for imaginary values of x_0 and φ by replacing φ by $i\varphi$, x_0 by ix_0 and x_0' by ix_0'. Instead of (22.19) we then have

$$ix'_0 = ix_0 \cos i\varphi - x_1 \sin i\varphi, \quad x'_1 = ix_0 \sin i\varphi + x_1 \cos i\varphi \quad (22.20)$$

or

$$x'_0 = x_0 \operatorname{ch} \varphi + x_1 \operatorname{sh} \varphi, \quad x'_1 = x_0 \operatorname{sh} \varphi + x_1 \operatorname{ch} \varphi \quad (22.21)$$

If the transformation (22. 19) is such that

$$x_0^2 + x_1^2 = \text{inv} \quad (22.22)$$

the invariant of the transformation (22. 21) is obtained from (22. 22) by replacing x_0 with ix_0 :

$$-x_0^2 + x_1^2 = \text{inv} \quad (22.23)$$

However, this quadratic form is the same as the invariant of the Lorentz transformation in the (x_0x_1)-plane. We therefore conclude that the transformation (22. 21) is a two-dimensional Lorentz transformation. The relation between Lorentz transformations and the transformations in the $O^+(4)$ group in the planes (x_0x_2) and (x_0x_3) can be found in a similar way. Finally, the transformations in the planes (x_1x_2), (x_2x_3), and (x_1x_3) are the same for both groups and form the group $O^+(3)$.

We thus see that the transformations in the Lorentz group are obtained from four-dimensional rotations by replacing real rotational parameters in the two-dimensional planes (x_0x_i) $(i = 1, 2, 3)$ by purely imaginary quantities and, at the same time, replacing x_0 by ix_0. Since the matrix elements of four-dimensional rotations are periodic functions, the one-to-one correspondence between the Lorentz group of transformations and the transformations in the group $O^+(4)$ exists only in a definite neighborhood of the unit element. If the four-dimensional rotation matrix is denoted by $O(\varphi_{01}, \varphi_{02}, \varphi_{03}; \psi_{12}, \psi_{13}, \psi_{23})$, the corresponding matrix of the Lorentz group can be written in the form

$$\Lambda(\varphi_{01}, \varphi_{02}, \varphi_{03}; \psi_{12}, \psi_{13}, \psi_{23})$$
$$= V^{-1}O(i\varphi_{01}, i\varphi_{02}, i\varphi_{03}; \psi_{12}, \psi_{13}, \psi_{23})V \quad (22.24)$$

where

$$V = \begin{Vmatrix} i & 0 & 0 & 0 \\ 0 & 1 & 0 & 0 \\ 0 & 0 & 1 & 0 \\ 0 & 0 & 0 & 1 \end{Vmatrix}$$

22.3 Commutation relations for infinitesimal matrices

We shall now derive the commutation relations for infinitesimal matrices in the Lorentz group, corresponding to rotations in two-dimensional planes. To do this, we shall first write the commutation relation for infinitesimal matrices in the group $O^+(4)$, and then use the formula given by (22.24) to find the commutation relations for the Lorentz group.

The four-dimensional rotation group includes four sub-groups of three-dimensional rotations which operate in the spaces R_{ikj} with coordinates x_i, x_k, x_j:

$$R_{012}: x_0, x_1, x_2$$
$$R_{123}: x_1, x_2, x_3$$
$$R_{230}: x_2, x_3, x_0$$
$$R_{301}: x_3, x_0, x_1$$

The infinitesimal matrices of the group $O^+(4)$ which we shall denote by $A_{01}, A_{02}, A_{03}, A_{12}, A_{13}, A_{23}$ $(A_{ik} = -A_{kl})$ can, at the same time, be regarded as the infinitesimal matrices of the following four groups of three-dimensional rotations

$$R_{012}: A_{12}, A_{20}, A_{01}$$
$$R_{123}: A_{23}, A_{31}, A_{12}$$
$$R_{230}: A_{30}, A_{02}, A_{23}$$
$$R_{301}: A_{01}, A_{13}, A_{30}$$

For each of the above triplets of infinitesimal matrices we have the following commutation relations (see Chapter 11)

$$A_i A_k - A_k A_i \equiv [A_i, A_k] = A_j \tag{22.25}$$

We thus obtain 12 commutation relations. The remaining three relations are

$$[A_{01}, A_{23}]=0, \quad [A_{12}, A_{30}]=0, \quad [A_{02}, A_{13}]=0 \qquad (22.26)$$

since the matrices acting on different pairs of coordinates must commute.

We can now readily find the commutation relations for the infinitesimal matrices of the Lorentz group. According to (22. 24), the relation between the infinitesimal matrices in the two groups is

$$\left.\begin{aligned}\left(\frac{\partial \Lambda}{\partial \varphi}\right)_0 &= iV^{-1}\left(\frac{\partial O}{\partial \varphi}\right)_0 V, \qquad \varphi=\varphi_{01}, \varphi_{02}, \varphi_{03} \\ \left(\frac{\partial \Lambda}{\partial \psi}\right)_0 &= V^{-1}\left(\frac{\partial O}{\partial \psi}\right)_0 V, \qquad \psi=\psi_{12}, \psi_{13}, \psi_{23}\end{aligned}\right\} \qquad (22.27)$$

The similarity transformation does not modify the commutation relations and, therefore, the commutation relations for infinitesimal matrices of the Lorentz groups B_{01}, B_{02}, B_{03}, B_{12}, B_{13}, and B_{23} are the same for the matrices iA_{01}, iA_{02}, iA_{03}, A_{12}, A_{13}, A_{23}. This property is, of course, also valid for the infinitesimal matrices of the representations of these groups; we shall use the notation B_{ik} and A_{ik} also for the infinitesimal matrices of the representations.

22.4 Irreducible representations

We already know that the determination of the irreducible representations of a continuous group can be reduced to the determination of the infinitesimal matrices of the representations. We shall now show that the determination of finite-dimensional irreducible representations of the Lorentz group can be reduced to the analogous problem for the group $O^+(4)$. In fact, if we know the infinitesimal matrices of the irreducible representations of the group $O^+(4)$, the corresponding infinitesimal matrices for the Lorentz group will either be the same (for space rotations) or differ from them by the factor i (in the case of transformations relating the time and the space coordinates).

To determine the infinitesimal matrices of the irreducible representations of the group $O^+(4)$, we shall proceed as follows. Consider the matrices

$$\left.\begin{aligned} \mu_1 &= \frac{1}{2}(A_{23}+A_{01}), & \tau_1 &= \frac{1}{2}(A_{23}-A_{01}) \\ \mu_2 &= \frac{1}{2}(A_{31}+A_{02}), & \tau_2 &= \frac{1}{2}(A_{31}-A_{02}) \\ \mu_3 &= \frac{1}{2}(A_{12}+A_{03}), & \tau_3 &= \frac{1}{2}(A_{12}-A_{03}) \end{aligned}\right\} \quad (22.28)$$

It is readily verified, using the commutation relations for the matrices A_{ik}, that the matrices $\boldsymbol{\mu}$ and $\boldsymbol{\tau}$ satisfy the same commutation relations as the infinitesimal matrices of the three-dimensional rotation group, i.e.

$$[\boldsymbol{\mu}, \boldsymbol{\mu}] = \boldsymbol{\mu}, \quad [\boldsymbol{\tau}, \boldsymbol{\tau}] = \boldsymbol{\tau} \quad (22.29)$$

Moreover, the matrices μ_k and τ_i commute. We can satisfy all these requirements if we take

$$\left.\begin{aligned} \mu_i &= M_i \times E_T \\ \tau_i &= E_M \times T_i \end{aligned}\right\} \quad (22.30)$$

for the matrices μ_k and τ_i, where M_i and T_i are the infinitesimal matrices of two representations of the three-dimensional group and E_T and E_M are unit matrices of these representations. From (22. 30) we may conclude that the four-dimensional rotation group is isomorphic to the product of two three-dimensional rotation groups. Since all the irreducible representations of the direct product of two groups can be formed by the composition of the irreducible representations of the two multiplicands (see Chapter 4), the irreducible representations of the group $O^+(4)$ form the composition of the irreducible representations of two three-dimensional rotation groups.

We know that each irreducible representation of the rotation group is unambiguously determined by its weight j, which can be equal to a non-negative integral or half-integral number. Irreducible representations of the group

$O^{+}(4)$, therefore, are defined by the two numbers j and j', each of which can be integral or half-integral. The order of such a representation is equal to the product of the orders of the irreducible representations of the three-dimensional rotation group, i.e.

$$(2j+1)(2j'+1)$$

The infinitesimal matrices of an irreducible representation of the group $O^{+}(4)$ will be determined by the formulae (22.30) with M_i and T_i replaced by the infinitesimal matrices of the irreducible representations of the group $O^{+}(3)$.

If we now introduce for the infinitesimal matrices of the group $O^{+}(4)$ the new notation

$$\left.\begin{aligned} A_1^{(+)} &= A_{23}, & A_1^{(-)} &= A_{01} \\ A_2^{(+)} &= A_{31}, & A_2^{(-)} &= A_{02} \\ A_3^{(+)} &= A_{12}, & A_3^{(-)} &= A_{03} \end{aligned}\right\} \tag{22.31}$$

we can write

$$A_i^{(\pm)} = (M_i \times E_T \pm E_M \times T_i) \tag{22.32}$$

For the infinitesimal matrices of irreducible representations of the Lorentz group we shall introduce the notation analogous to (22.31), so that the final expressions are

$$\left.\begin{aligned} B_i^{(+)} &= (M_i \times E_T + E_M \times T_i) \\ B_i^{(-)} &= i(M_i \times E_T - E_M \times T_i) \end{aligned}\right\} \tag{22.33}$$

The irreducible representations of the proper Lorentz group found in this way are thus determined by the pair of numbers j and j', each of which can be integral or half-integral. We shall denote these representations by $D^{(jj')}$. The order of the representation $D^{(jj')}$ is $(2j+1)(2j'+1)$. The infinitesimal matrices of this representation are given by (22.33) in which M_i and T_i are the infinitesimal matrices of the irreducible representations of the rotation group with weights j and j', respectively.

We note that, owing to the presence of the factor i in

the second formula in (22.33), the infinitesimal matrices $B_i^{(-)}$ are not anti-Hermitian and, consequently, the representations $D^{(jj')}$ are not unitary. The non-unitarity of finite-dimensional representations of the Lorentz group is connected with the fact that this group is not compact. In addition to these finite-dimensional irreducible representations the Lorentz group has infinite-dimensional unitary irreducible representations. These representations have a restricted range of application in quantum mechanics (Chapter 14) and we shall not consider them here.

22.5 Direct product of irreducible representations of the Lorentz group

To consider this problem, we shall again use the correspondence between the representations of the Lorentz group and those of the group $O^+(4)$. We shall denote the irreducible representations of the latter group, which correspond to the irreducible representations $D^{(jj')}$ of the Lorentz group, by $\tilde{D}^{(jj')}$.

We have seen that the irreducible representations of the group $O^+(4)$ can be expressed in the form of the direct product of the irreducible representations of two three-dimensional rotation groups:

$$\tilde{D}^{(jj')} = D^{(j)}(g) \times D^{(j')}(g') \tag{22.34}$$

where g and g' are the elements belonging to the two independent three-dimensional rotation groups. Let us find the decomposition of the direct product $\tilde{D}^{(j_1 j_1')} \times \tilde{D}^{(j_2 j_2')}$ into irreducible parts. According to (22.34), we can write

$$\tilde{D}^{(j_1 j_1')} \times \tilde{D}^{(j_2 j_2')} = D^{(j_1)}(g) \times D^{(j_1')}(g') \times D^{(j_2)}(g) \times D^{(j_2')}(g') \tag{22.35}$$

If we consider that the representation matrices are equal to within the similarity transformation, then on the right-hand side of (22.35) we can change the order of the factors, and if we use the Clebsch–Gordan rule for the $O^+(3)$ groups (see Chapter 12), we obtain

$$\tilde{D}^{(j_1 j_1')} \times \tilde{D}^{(j_2 j_2')}$$
$$= (D^{(j_1)}(g) \times D^{(j_2)}(g)) \times \left(D^{(j_1')}(g') \times D^{(j_2')}(g')\right)$$
$$= \sum_{j=|j_1-j_2|}^{j_1+j_2}{}^{\oplus} D^{(j)}(g) \times \sum_{j'=|j_1'-j_2'|}^{j_1'+j_2'}{}^{\oplus} D^{(j')}(g') \qquad (22.36)$$

According to (22.34), this result can be reduced to the form

$$\tilde{D}^{(j_1 j_1')} \times \tilde{D}^{(j_2 j_2')} = \sum_{j=|j_1-j_2|}^{j_1+j_2} \sum_{j'=|j_1'-j_2'|}^{j_1'+j_2'}{}^{\oplus} \tilde{D}^{(jj')} \qquad (22.37)$$

It is clear that, because of the correspondence between the irreducible representations of the Lorentz and $O^+(4)$ groups, we obtain a similar decomposition for the direct product of two irreducible finite-dimensional representations of the Lorentz group.

Exercises

22.1. Show that an arbitrary proper Lorentz transformation can be written as the product of six one-parameter transformations, each in a certain plane.

22.2. Write down the explicit expression for the infinitesimal matrices of the following irreducible representations:

$$D^{\left(0\,\frac{1}{2}\right)},\quad D^{\left(\frac{1}{2}\,0\right)},\quad D^{\left(\frac{1}{2}\,\frac{1}{2}\right)},\quad D^{\left(\frac{1}{2}\,\frac{3}{2}\right)}$$

22.3. Find the irreducible representations of the three-dimensional rotation sub-group into which the representation $D^{(jj')}$ can be decomposed.

22.4. Find the decomposition into irreducible representations of

$$D^{\left(\frac{1}{2}\,0\right)} \times D^{\left(\frac{1}{2}\,\frac{1}{2}\right)},\quad D^{\left(\frac{1}{2}\,\frac{1}{2}\right)} \times D^{\left(\frac{1}{2}\,\frac{1}{2}\right)}$$

22.5. Show that the four-dimensional vector (x_0, x_1, x_2, x_3) transforms in accordance with the irreducible representation $D^{\left(\frac{1}{2}\,\frac{1}{2}\right)}$.

Chapter 23

The Dirac Equation

As an illustration of the application of the theory of representations of the Lorentz group we shall consider the relativistically invariant wave equation for a free particle with spin 1/2, i.e. the Dirac equation. Since, in this case, the wave function is a multi-component quantity which must transform in accordance with a representation of the Lorentz group, this equation is, in fact, a set of linear differential equations. We shall suppose that the reader is familiar with the Dirac equation as introduced in courses on quantum mechanics. We shall therefore confine ourselves to the transformation properties of its solution.

23.1 Relativistically invariant equations

The set of linear first-order differential equations describing the state of a free particle can be written in the form

$$L_0 \frac{\partial \psi}{\partial x_0} + L_1 \frac{\partial \psi}{\partial x_1} + L_2 \frac{\partial \psi}{\partial x_2} + L_3 \frac{\partial \psi}{\partial x_3} + i\varkappa\psi = 0 \qquad (23.1)$$

where $\psi = \begin{pmatrix} \psi_1 \\ \psi_2 \\ \cdots \end{pmatrix}$ is a multi-component wave function of the variables x_0, x_1, x_2, x_3, and $L_0, \ldots, L_3$ and $\varkappa$ are certain matrices whose elements do not depend on the coordinates or the time.

Let X and $X' = \Lambda X$ represent the same space-time point in two reference frames. The transformation rule for the wave function between the primed and unprimed sets of coordinates is

$$\psi'(x') = D(\Lambda)\,\psi(x) \tag{23.2}$$

or

$$\psi'_\alpha(x') = \sum_\beta D_{\alpha\beta}(\Lambda)\,\psi_\beta(x)$$

We shall show that the matrices $D(\Lambda)$ form a representation of the Lorentz group. In fact, the successive application of two Lorentz transformations yields

$$\psi''_\gamma(x'') = \sum_\delta D_{\gamma\delta}(\Lambda')\,\psi'_\delta(x')$$

$$= \sum_{\beta,\,\delta} D_{\gamma\delta}(\Lambda')\,D_{\delta\beta}(\Lambda)\,\psi_\beta(x) = \sum_\beta D_{\gamma\beta}(\Lambda'')\,\psi_\beta(x) \tag{23.3}$$

If we write (23. 1) in the primed set of coordinates, we obtain, after applying the reverse transformation,

$$\psi(x) = D^{-1}(\Lambda)\,\psi'(x') \tag{23.4}$$

Next, we have

$$x'_i = \sum_{k=0}^{3} \Lambda_{ik} x_k \tag{23.5}$$

and, consequently,

$$\frac{\partial}{\partial x_i} = \sum_{k=0}^{3} \Lambda_{ki} \frac{\partial}{\partial x'_k} \tag{23.6}$$

Substituting (23. 4) and (23. 6) into (23. 1), we obtain

$$\sum_{i,\,k=0}^{3} L_i D^{-1} \frac{\partial \psi'}{\partial x'_k} \Lambda_{ki} + i\varkappa D^{-1}\psi' = 0 \tag{23.7}$$

The matrix $\varkappa$ will be assumed to be non-singular, and, we shall assume it to be equal to a multiple of the unit matrix.

Consequently, we can always multiply both sides of (23. 1) by $\varkappa^{-1}$. Thus, multiplying (23. 7) on the left by D, we obtain

$$\sum_{k,\, i=0}^{3} \Lambda_{ki} D L_i D^{-1} \frac{\partial \psi'}{\partial x_k'} + i\varkappa\psi' = 0 \tag{23.8}$$

Since (23. 1) must be invariant under the Lorentz transformation, we have

$$L_k \doteq \sum_{i=0}^{3} \Lambda_{ki} D L_i D^{-1} \tag{23.9}$$

or

$$D^{-1} L_k D = \sum_{i=0}^{3} \Lambda_{ki} L_i \tag{23.10}$$

This shows that the matrices L_k transform like a four-dimensional vector. The formulae given by (23. 9) express the invariance of the above system of differential equations.

We shall now determine the form of the simplest relativistically invariant systems (23. 1) consisting of the minimum number of equations. Since the number of equations is equal to the number of components of the wave function, and the latter transforms in accordance with a representation of the Lorentz group, it is natural to begin with the equation corresponding to an irreducible representation of the Lorentz group with minimum dimensionality. We can begin with the one-dimensional identity representation $D^{(00)}$. However, it can be verified that a relativistically invariant equation of the form of (23. 1) cannot be written for this representation. In point of fact, if we transfer the term $i\varkappa\psi$ in (23. 1) to the right-hand side of the equation, we find that the left-hand side includes a quantity which transforms by the representation $D^{(00)} \times D^{\left(\frac{1}{2}\frac{1}{2}\right)}$, whereas the right-hand side transforms by the representation $D^{(00)}$ (see Exercise 22. 5). It is clear that two quantities transforming by different representations cannot be equal. In order to be able to write down a relativistically invariant system whose solution transforms by a representation containing the identity representation, the former must be augmented at least by the representation $D^{(\frac{1}{2}\,\frac{1}{2})}$. Under these conditions, we would obtain a system consisting of five equations. Let us now consider the equation for the wave function which transforms by the representation $D^{(\frac{1}{2}\,0)}$ of order 2. Proceeding as above, we see that the left-hand side of the equation contains a quantity which transforms in accordance with the representation

$$D^{(\frac{1}{2}\,0)} \times D^{(\frac{1}{2}\,\frac{1}{2})} = D^{(1\,\frac{1}{2})} \oplus D^{(0\,\frac{1}{2})}$$

whereas the right-hand side transforms in accordance with the representation

$$D^{(\frac{1}{2}\,0)}$$

i.e. here again, we cannot write down a relativistically invariant system. However, we shall not reach a contradiction if we suppose that the wave function transforms in accordance with the irreducible representation $D^{(\frac{1}{2}\,0)} \oplus D^{(0\,\frac{1}{2})}$ of order 4. Substituting $D = D^{(\frac{1}{2}\,0)} \oplus D^{(0\,\frac{1}{2})}$ in (23. 10), we can unambiguously determine the matrices L_i. The resulting system will consist of four equations for the four components of the wave function. This system, written in the form of (23. 1), is called the Dirac equation.

23.2 The Dirac equation

The matrices L_i can be found from (23. 10). The explicit form of the Dirac equation is

$$L_0 \frac{\partial \psi}{\partial x_0} + L_1 \frac{\partial \psi}{\partial x_1} + L_2 \frac{\partial \psi}{\partial x_2} + L_3 \frac{\partial \psi}{\partial x_3} + i\varkappa\psi = 0 \qquad (23.11)$$

where

$$\left.\begin{aligned}
L_0 &= \begin{pmatrix} 0 & 0 & 1 & 0 \\ 0 & 0 & 0 & 1 \\ 1 & 0 & 0 & 0 \\ 0 & 1 & 0 & 0 \end{pmatrix}, \quad L_1 = \begin{pmatrix} 0 & 0 & 0 & -1 \\ 0 & 0 & -1 & 0 \\ 0 & 1 & 0 & 0 \\ 1 & 0 & 0 & 0 \end{pmatrix} \\
L_2 &= \begin{pmatrix} 0 & 0 & 0 & i \\ 0 & 0 & -i & 0 \\ 0 & -i & 0 & 0 \\ i & 0 & 0 & 0 \end{pmatrix}, \quad L_3 = \begin{pmatrix} 0 & 0 & 1 & 0 \\ 0 & 0 & 0 & -1 \\ -1 & 0 & 0 & 0 \\ 0 & 1 & 0 & 0 \end{pmatrix}
\end{aligned}\right\} \tag{23.12}$$

The four-component wave function (the Dirac bispinor) $\psi = \begin{pmatrix} \psi_1 \\ \psi_2 \\ \psi_3 \\ \psi_4 \end{pmatrix}$ transforms in accordance with the representation

$$D = D^{(\frac{1}{2}\,0)} \oplus D^{(0\,\frac{1}{2})}$$

i.e. the two components ψ_1, ψ_2 transform by the representation $D^{(\frac{1}{2}\,0)}$, whereas ψ_3, ψ_4 transform by the representation $D^{(0\,\frac{1}{2})}$. Using the result obtained in the last chapter, we find that the infinitesimal operators of the representation $D^{(\frac{1}{2}\,0)}$ are given by

$$\left.\begin{aligned}
B_1^{(+)} &= \frac{i}{2}\begin{pmatrix} 0 & 1 \\ 1 & 0 \end{pmatrix}, \quad B_1^{(-)} = -\frac{1}{2}\begin{pmatrix} 0 & 1 \\ 1 & 0 \end{pmatrix} \\
B_2^{(+)} &= \frac{1}{2}\begin{pmatrix} 0 & -1 \\ 1 & 0 \end{pmatrix}, \quad B_2^{(-)} = \frac{i}{2}\begin{pmatrix} 0 & -1 \\ 1 & 0 \end{pmatrix} \\
B_3^{(+)} &= \frac{i}{2}\begin{pmatrix} 1 & 0 \\ 0 & -1 \end{pmatrix}, \quad B_3^{(-)} = -\frac{1}{2}\begin{pmatrix} 1 & 0 \\ 0 & -1 \end{pmatrix}
\end{aligned}\right\} \tag{23.13}$$

whereas the infinitesimal operators of the representation $D^{(0\,\frac{1}{2})}$ are given by

$$\left.\begin{aligned}
B_1^{(+)} &= \frac{i}{2}\begin{pmatrix} 0 & 1 \\ 1 & 0 \end{pmatrix}, \quad B_1^{(-)} = \frac{1}{2}\begin{pmatrix} 0 & 1 \\ 1 & 0 \end{pmatrix} \\
B_2^{(+)} &= \frac{1}{2}\begin{pmatrix} 0 & -1 \\ 1 & 0 \end{pmatrix}, \quad B_2^{(-)} = -\frac{i}{2}\begin{pmatrix} 0 & -1 \\ 1 & 0 \end{pmatrix} \\
B_3^{(+)} &= \frac{i}{2}\begin{pmatrix} 1 & 0 \\ 0 & -1 \end{pmatrix}, \quad B_3^{(-)} = \frac{1}{2}\begin{pmatrix} 1 & 0 \\ 0 & -1 \end{pmatrix}
\end{aligned}\right\} \tag{23.14}$$

If instead of the 'canonical' components of the wave function we take their linear combinations $\varphi_\alpha = \sum C_{\alpha\beta}\psi_\beta$, the equation for the function φ will be of the form (23. 11) with the matrices $L_i' = CL_iC^{-1}$. If, for example, we substitute

$$\left.\begin{aligned} \varphi_1 &= \psi_2 + \psi_4, & \varphi_3 &= -\psi_2 + \psi_4 \\ \varphi_2 &= -(\psi_1 + \psi_3), & \varphi_4 &= \psi_1 - \psi_3 \end{aligned}\right\} \tag{23.15}$$

we obtain an equation for the matrices $L_i' = \gamma_i$, which have the following explicit form

$$\left.\begin{aligned} \gamma_0 &= \begin{pmatrix} 1 & 0 & 0 & 0 \\ 0 & 1 & 0 & 0 \\ 0 & 0 & -1 & 0 \\ 0 & 0 & 0 & -1 \end{pmatrix}, & \gamma_1 &= \begin{pmatrix} 0 & 0 & 0 & 1 \\ 0 & 0 & 1 & 0 \\ 0 & -1 & 0 & 0 \\ -1 & 0 & 0 & 0 \end{pmatrix} \\ \gamma_2 &= \begin{pmatrix} 0 & 0 & 0 & -i \\ 0 & 0 & i & 0 \\ 0 & i & 0 & 0 \\ -i & 0 & 0 & 0 \end{pmatrix}, & \gamma_3 &= \begin{pmatrix} 0 & 0 & 1 & 0 \\ 0 & 0 & 0 & -1 \\ -1 & 0 & 0 & 0 \\ 0 & 1 & 0 & 0 \end{pmatrix} \end{aligned}\right\} \tag{23.16}$$

This is, in fact, the more common way of expressing the Dirac equation.

23.3 The complex conjugate Dirac bispinor

We shall now find the transformation rule for the complex conjugate wave function $\bar{\psi}$. It is clear that complex conjugate functions transform by representations with complex conjugate matrices.

If we extend our analysis somewhat, we can formulate the problem in the following way. Suppose that we have an irreducible representation $D^{(j_1 j_2)}$ of the Lorentz group. If instead of each matrix of this representation we take the complex conjugate matrix, we again obtain a representation of the group which we shall call the complex conjugate representation $\bar{D}^{(j_1 j_2)}$. It is clear that this representation is also irreducible, i. e.

$$\overline{D}^{(j_1 j_2)} = D^{(j'_1 j'_2)}$$

It is required to find j'_1, j'_2. We recall that the infinitesimal matrices of the Lorentz group can be written

$$\left.\begin{aligned} B_l^{(+)} &= \frac{1}{2}(M_l \times E_T + E_M \times T_l) \\ B_l^{(-)} &= \frac{i}{2}(M_l \times E_T - E_M \times T_l) \end{aligned}\right\} \tag{23.17}$$

where T_l and M_l are the infinitesimal matrices of the irreducible representations of the group $O^+(3)$ with weights j_1 and j_2, respectively, and E_M and E_T are the unit matrices of these representations. By taking the complex conjugates, we obtain

$$\overline{B}_l^{(+)} = \frac{1}{2}(\overline{M}_l \times E_T + E_M \times \overline{T}_l) \tag{23.18}$$

$$\overline{B}_l^{(-)} = \frac{i}{2}(E_M \times \overline{T}_l - \overline{M}_l \times E_T) \tag{23.19}$$

We know that the interchange of the factors in the direct product of matrices is equivalent to a similarity transformation, and we can therefore write

$$\overline{B}_l^{(+)} = \frac{1}{2} U [E_T \times \overline{M}_l + \overline{T}_l \times E_M] U^{-1} \tag{23.20}$$

$$\overline{B}_l^{(-)} = \frac{i}{2} U [\overline{T}_l \times E_M - E_T \times \overline{M}_l] U^{-1} \tag{23.21}$$

Next, we know (see Section 12.4) that the matrices $\overline{M}_l$ and $\overline{T}_l$ can be expressed in terms of the canonical matrices M_l and T_l:

$$V_T T_l V_T^{-1} = \overline{T}_l, \quad V_M M_l V_M^{-1} = \overline{M}_l \tag{23.22}$$

where V_T and V_M are the matrices of the corresponding representations for a rotation through 180° about the y-axis. If we now substitute (23.22) in (23.20) and (23.21), we obtain

$$\left.\begin{aligned} \overline{B}_l^{(+)} &= \frac{1}{2} W [E_T \times M_l + T_l \times E_M] W^{-1} \\ \overline{B}_l^{(-)} &= \frac{i}{2} W [T_l \times E_M - E_T \times M_l] W^{-1} \end{aligned}\right\} \tag{23.23}$$

where

$$W = (E_T \times V_M)(V_T \times E_M) U \tag{23.24}$$

Comparison of (23. 23) with (23. 17) will show that the complex conjugate infinitesimal matrices $\overline{B}_l^{(+)}$ and $\overline{B}_l^{(-)}$ coincide to within the similarity transformation with the canonical infinitesimal operators of the irreducible representation $D^{(j_2 j_1)}$. Thus, we find that

$$j_1' = j_2, \quad j_2' = j_1 \tag{23.25}$$

However, in addition to this result we have also found the explicit form of the transformation which takes the complex conjugate infinitesimal operators into the canonical form. This transformation is given by (23. 24). Thus, if $\{q_\alpha\}$ is the basis for the representation $D^{(j_1 j_2)}$, then $\{\overline{q}_\alpha\}$ is a basis for $D^{(j_2 j_1)}$. The basis $\{\overline{q}_\alpha\}$ is not canonical, and transition to the canonical basis is achieved by the transformation

$$\tilde{q}_\alpha = \sum_\beta W_{\beta\alpha} \overline{q}_\beta \tag{23.26}$$

Let us write this transformation for the representation $D^{\left(0\ \frac{1}{2}\right)}$. Since the basis of this representation is also the basis of an irreducible representation of the rotation group, then from (12. 52) we have

$$\tilde{q} = (2A_2)^{-1} \overline{q} \tag{23.27}$$

or

$$\tilde{q}_{\frac{1}{2}} = \overline{q}_{-\frac{1}{2}}, \quad \tilde{q}_{-\frac{1}{2}} = -\overline{q}_{\frac{1}{2}}$$

We have shown that the quantities $\tilde{q}_{\frac{1}{2}}, \tilde{q}_{-\frac{1}{2}}$ form the basis of the representation $D^{\left(\frac{1}{2}\ 0\right)}$. Conversely, if $p_{\frac{1}{2}}, p_{-\frac{1}{2}}$ is the basis of the representation $D^{\left(\frac{1}{2}\ 0\right)}$, then it follows from (23. 27) that the quantities

$$\tilde{p}_{\frac{1}{2}} = -\overline{p}_{-\frac{1}{2}}, \quad \tilde{p}_{-\frac{1}{2}} = \overline{p}_{\frac{1}{2}} \tag{23.28}$$

form the basis of the irreducible representation $D^{\left(0\ \frac{1}{2}\right)}$ These results can be used directly to find the transformation rule for the complex conjugate Dirac bispinor. We have

$$\bar{\psi}_1 = \psi'_4, \quad \bar{\psi}_2 = -\psi'_3, \quad \bar{\psi}_3 = \psi'_2, \quad \bar{\psi}_4 = -\psi'_1 \tag{23.29}$$

where $\psi'_1, \psi'_2, \psi'_3, \psi'_4$ are the four components of a new bispinor.

23.4 The invariant quadratic form

Let us now use the components of the bispinor to construct the Hermitian quadratic form which is invariant under proper Lorentz transformations

$$\text{inv} = \sum_{i,k} \bar{\psi}_i \psi_k a_{ik} = \sum_{i,k} \psi'_i \psi_k b_{ik} \tag{23.30}$$

To find the coefficients a_{ik} we note that:

1. The identity representation of the Lorentz group is contained only in the direct product of identical representations.

2. The quadratic form must also be invariant under the rotation group.

3. $a_{ik} = \bar{a}_{ki}$ (the condition of Hermiticity).

From the first two conditions we have

$$\begin{aligned} \text{inv} &= a(\psi_1\psi'_1 - \psi_2\psi'_2) + b(\psi_3\psi'_3 - \psi_4\psi'_4) \\ &= -a(\psi_1\bar{\psi}_4 + \psi_2\bar{\psi}_3) + b(-\psi_3\bar{\psi}_2 - \psi_4\bar{\psi}_1) \end{aligned} \tag{23.31}$$

where a and b are arbitrary complex numbers. From the third condition we have $a = \bar{b}$ and $a_{ik} = \bar{a}_{ki}$

$$\text{inv} = a(\psi_1\bar{\psi}_4 + \psi_2\bar{\psi}_3) + \bar{a}(\psi_3\bar{\psi}_2 + \psi_4\bar{\psi}_1) \tag{23.32}$$

We then readily verify that this form is not positive-definite, which is in agreement with the fact that the representations of the Lorentz group which we are considering are not unitary. In particular, if we set $a = 1$, we have

$$\text{inv} = \tilde{\psi}\psi \tag{23.33}$$

where $\tilde{\psi}=(\bar{\psi}_4, \bar{\psi}_3, \bar{\psi}_2, \bar{\psi}_1)$ is a row matrix and $\psi=\begin{Bmatrix}\psi_1\\ \psi_2\\ \psi_3\\ \psi_4\end{Bmatrix}$ is a column matrix. The transformation rules for ψ and $\tilde{\psi}$ under the Lorentz transformation can be written in the form

$$\psi'=D\psi, \quad \tilde{\psi}'=\tilde{\psi}\tilde{D} \tag{23.34}$$

Since $\tilde{\psi}\psi$ is an invariant, it follows that

$$\tilde{\psi}'\psi'=\tilde{\psi}\tilde{D}D\psi=\tilde{\psi}\psi \tag{23.35}$$

and hence

$$\tilde{D}=D^{-1}$$

Using this result and Equation (23.10), we can readily show that, for example, the quantities $\tilde{\psi}L_i\psi$ transform like a four-dimensional vector.

Appendix to Chapter 7

Matrices of the irreducible representations of the group O

	E	$C_2^{(x)}$	$C_2^{(y)}$	$C_2^{(z)}$	$C_4^{(x)}$	$C_4^{(y)}$	$C_4^{(z)}$	$(C_4^{(x)})^3$
$\Gamma^{(1)}$	1	1	1	1	1	1	1	1
$\Gamma^{(2)}$	1	1	1	1	-1	-1	-1	-1
$\Gamma^{(3)}$ x^2-y^2 $3z^2-r^2$	$\begin{pmatrix}1&0\\0&1\end{pmatrix}$	$\begin{pmatrix}1&0\\0&1\end{pmatrix}$	$\begin{pmatrix}1&0\\0&1\end{pmatrix}$	$\begin{pmatrix}1&0\\0&1\end{pmatrix}$	$\begin{pmatrix}-1&0\\0&1\end{pmatrix}$	$\begin{pmatrix}\frac{1}{2}&\frac{\sqrt{3}}{2}\\\frac{\sqrt{3}}{2}&-\frac{1}{2}\end{pmatrix}$	$\begin{pmatrix}\frac{1}{2}&-\frac{\sqrt{3}}{2}\\-\frac{\sqrt{3}}{2}&-\frac{1}{2}\end{pmatrix}$	$\begin{pmatrix}-1&0\\0&1\end{pmatrix}$
$\Gamma^{(4)}$	$\begin{pmatrix}1&0&0\\0&1&0\\0&0&1\end{pmatrix}$	$\begin{pmatrix}1&0&0\\0&-1&0\\0&0&-1\end{pmatrix}$	$\begin{pmatrix}-1&0&0\\0&1&0\\0&0&-1\end{pmatrix}$	$\begin{pmatrix}-1&0&0\\0&-1&0\\0&0&1\end{pmatrix}$	$\begin{pmatrix}-1&0&0\\0&0&1\\0&-1&0\end{pmatrix}$	$\begin{pmatrix}0&0&-1\\0&-1&0\\1&0&0\end{pmatrix}$	$\begin{pmatrix}0&1&0\\-1&0&0\\0&0&-1\end{pmatrix}$	$\begin{pmatrix}-1&0&0\\0&0&-1\\0&1&0\end{pmatrix}$
$\Gamma^{(5)}$ x, y, z	$\begin{pmatrix}1&0&0\\0&1&0\\0&0&1\end{pmatrix}$	$\begin{pmatrix}1&0&0\\0&-1&0\\0&0&-1\end{pmatrix}$	$\begin{pmatrix}-1&0&0\\0&1&0\\0&0&-1\end{pmatrix}$	$\begin{pmatrix}-1&0&0\\0&-1&0\\0&0&1\end{pmatrix}$	$\begin{pmatrix}1&0&0\\0&0&-1\\0&1&0\end{pmatrix}$	$\begin{pmatrix}0&0&1\\0&1&0\\-1&0&0\end{pmatrix}$	$\begin{pmatrix}0&-1&0\\1&0&0\\0&0&1\end{pmatrix}$	$\begin{pmatrix}1&0&0\\0&0&1\\0&-1&0\end{pmatrix}$

Notation. $C_k^{(a)}$ is a k-fold axis passing through the point a (see Fig. 18).

	$C_3^{(a)}$	$C_3^{(b)}$	$C_3^{(c)}$	$C_3^{(d)}$	$(C_3^{(a)})^2$	$(C_3^{(b)})^2$	$(C_3^{(c)})^2$	$(C_3^{(d)})^2$
$\Gamma^{(1)}$	1	1	1	1	1	1	1	1
$\Gamma^{(2)}$	1	1	1	1	1	1	1	1
$\Gamma^{(3)}$ x^2-y^2 $3z^2-r^2$	$\begin{pmatrix} -\frac{1}{2} & \frac{\sqrt{3}}{2} \\ -\frac{\sqrt{3}}{2} & -\frac{1}{2} \end{pmatrix}$	$\begin{pmatrix} -\frac{1}{2} & -\frac{\sqrt{3}}{2} \\ \frac{\sqrt{3}}{2} & -\frac{1}{2} \end{pmatrix}$	$\begin{pmatrix} -\frac{1}{2} & \frac{\sqrt{3}}{2} \\ -\frac{\sqrt{3}}{2} & -\frac{1}{2} \end{pmatrix}$	$\begin{pmatrix} -\frac{1}{2} & -\frac{\sqrt{3}}{2} \\ \frac{\sqrt{3}}{2} & -\frac{1}{2} \end{pmatrix}$	$\begin{pmatrix} -\frac{1}{2} & -\frac{\sqrt{3}}{2} \\ \frac{\sqrt{3}}{2} & -\frac{1}{2} \end{pmatrix}$	$\begin{pmatrix} -\frac{1}{2} & \frac{\sqrt{3}}{2} \\ -\frac{\sqrt{3}}{2} & -\frac{1}{2} \end{pmatrix}$	$\begin{pmatrix} -\frac{1}{2} & -\frac{\sqrt{3}}{2} \\ \frac{\sqrt{3}}{2} & -\frac{1}{2} \end{pmatrix}$	$\begin{pmatrix} -\frac{1}{2} & \frac{\sqrt{3}}{2} \\ -\frac{\sqrt{3}}{2} & -\frac{1}{2} \end{pmatrix}$
$\Gamma^{(4)}$	$\begin{pmatrix} 0 & 0 & -1 \\ 1 & 0 & 0 \\ 0 & -1 & 0 \end{pmatrix}$	$\begin{pmatrix} 0 & -1 & 0 \\ 0 & 0 & -1 \\ 1 & 0 & 0 \end{pmatrix}$	$\begin{pmatrix} 0 & 0 & 1 \\ 1 & 0 & 0 \\ 0 & 1 & 0 \end{pmatrix}$	$\begin{pmatrix} 0 & -1 & 0 \\ 0 & 0 & 1 \\ -1 & 0 & 0 \end{pmatrix}$	$\begin{pmatrix} 0 & 1 & 0 \\ 0 & 0 & -1 \\ -1 & 0 & 0 \end{pmatrix}$	$\begin{pmatrix} 0 & 0 & 1 \\ -1 & 0 & 0 \\ 0 & -1 & 0 \end{pmatrix}$	$\begin{pmatrix} 0 & 1 & 0 \\ 0 & 0 & 1 \\ 1 & 0 & 0 \end{pmatrix}$	$\begin{pmatrix} 0 & 0 & -1 \\ -1 & 0 & 0 \\ 0 & 1 & 0 \end{pmatrix}$
$\Gamma^{(5)}$ x, y, z	$\begin{pmatrix} 0 & 0 & -1 \\ 1 & 0 & 0 \\ 0 & -1 & 0 \end{pmatrix}$	$\begin{pmatrix} 0 & -1 & 0 \\ 0 & 0 & -1 \\ 1 & 0 & 0 \end{pmatrix}$	$\begin{pmatrix} 0 & 0 & 1 \\ 1 & 0 & 0 \\ 0 & 1 & 0 \end{pmatrix}$	$\begin{pmatrix} 0 & -1 & 0 \\ 0 & 0 & 1 \\ -1 & 0 & 0 \end{pmatrix}$	$\begin{pmatrix} 0 & 1 & 0 \\ 0 & 0 & -1 \\ -1 & 0 & 0 \end{pmatrix}$	$\begin{pmatrix} 0 & 0 & 1 \\ -1 & 0 & 0 \\ 0 & -1 & 0 \end{pmatrix}$	$\begin{pmatrix} 0 & 1 & 0 \\ 0 & 0 & 1 \\ 1 & 0 & 0 \end{pmatrix}$	$\begin{pmatrix} 0 & 0 & -1 \\ -1 & 0 & 0 \\ 0 & 1 & 0 \end{pmatrix}$

	$(C_4^{(y)})^3$	$(C_4^{(z)})^3$	$C_2^{(1)}$	$C_2^{(2)}$	$C_2^{(3)}$	$C_2^{(4)}$	$C_2^{(5)}$	$C_2^{(6)}$
$\Gamma^{(1)}$	1	1	1	1	1	1	1	1
$\Gamma^{(2)}$	−1	−1	−1	−1	−1	−1	−1	−1
$\Gamma^{(3)}$ x^2-y^2 $3z^2-r^2$	$\begin{pmatrix} \frac{1}{2} & \frac{\sqrt{3}}{2} \\ \frac{\sqrt{3}}{2} & -\frac{1}{2} \end{pmatrix}$	$\begin{pmatrix} \frac{1}{2} & -\frac{\sqrt{3}}{2} \\ -\frac{\sqrt{3}}{2} & -\frac{1}{2} \end{pmatrix}$	$\begin{pmatrix} \frac{1}{2} & -\frac{\sqrt{3}}{2} \\ \frac{\sqrt{3}}{2} & -\frac{1}{2} \end{pmatrix}$	$\begin{pmatrix} \frac{1}{2} & -\frac{\sqrt{3}}{2} \\ \frac{\sqrt{3}}{2} & -\frac{1}{2} \end{pmatrix}$	$\begin{pmatrix} \frac{1}{2} & \frac{\sqrt{3}}{2} \\ \frac{\sqrt{3}}{2} & -\frac{1}{2} \end{pmatrix}$	$\begin{pmatrix} \frac{1}{2} & \frac{\sqrt{3}}{2} \\ \frac{\sqrt{3}}{2} & -\frac{1}{2} \end{pmatrix}$	$\begin{pmatrix} -1 & 0 \\ 0 & 1 \end{pmatrix}$	$\begin{pmatrix} -1 & 0 \\ 0 & 1 \end{pmatrix}$
$\Gamma^{(4)}$	$\begin{pmatrix} 0 & 0 & 1 \\ 0 & -1 & 0 \\ -1 & 0 & 0 \end{pmatrix}$	$\begin{pmatrix} 0 & -1 & 0 \\ 1 & 0 & 0 \\ 0 & 0 & -1 \end{pmatrix}$	$\begin{pmatrix} 0 & 1 & 0 \\ 1 & 0 & 0 \\ 0 & 0 & 1 \end{pmatrix}$	$\begin{pmatrix} 0 & -1 & 0 \\ -1 & 0 & 0 \\ 0 & 0 & 1 \end{pmatrix}$	$\begin{pmatrix} 0 & 0 & 1 \\ 0 & 1 & 0 \\ 1 & 0 & 0 \end{pmatrix}$	$\begin{pmatrix} 0 & 0 & -1 \\ 0 & 1 & 0 \\ -1 & 0 & 0 \end{pmatrix}$	$\begin{pmatrix} 1 & 0 & 0 \\ 0 & 0 & 1 \\ 0 & 1 & 0 \end{pmatrix}$	$\begin{pmatrix} 1 & 0 & 0 \\ 0 & 0 & -1 \\ 0 & -1 & 0 \end{pmatrix}$
$\Gamma^{(5)}$ x, y, z	$\begin{pmatrix} 0 & 0 & -1 \\ 0 & 1 & 0 \\ 1 & 0 & 0 \end{pmatrix}$	$\begin{pmatrix} 0 & 1 & 0 \\ -1 & 0 & 0 \\ 0 & 0 & 1 \end{pmatrix}$	$\begin{pmatrix} 0 & -1 & 0 \\ -1 & 0 & 0 \\ 0 & 0 & -1 \end{pmatrix}$	$\begin{pmatrix} 0 & 1 & 0 \\ 1 & 0 & 0 \\ 0 & 0 & -1 \end{pmatrix}$	$\begin{pmatrix} 0 & 0 & -1 \\ 0 & -1 & 0 \\ -1 & 0 & 0 \end{pmatrix}$	$\begin{pmatrix} 0 & 0 & 1 \\ 0 & -1 & 0 \\ 1 & 0 & 0 \end{pmatrix}$	$\begin{pmatrix} -1 & 0 & 0 \\ 0 & 0 & -1 \\ 0 & -1 & 0 \end{pmatrix}$	$\begin{pmatrix} -1 & 0 & 0 \\ 0 & 0 & 1 \\ 0 & 1 & 0 \end{pmatrix}$

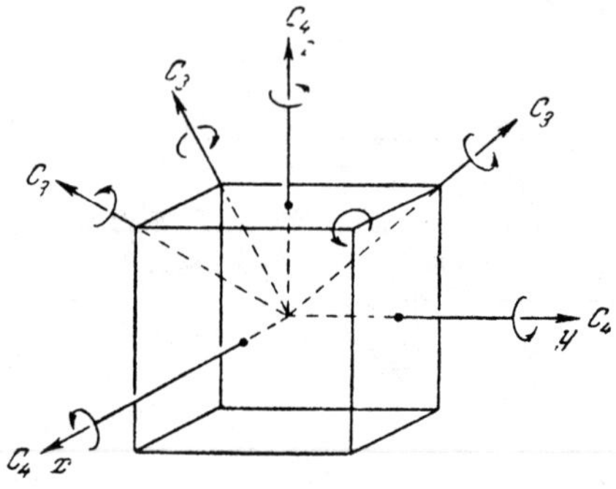

Fig. 18

Bibliography

1. WIGNER, E. P., Group theory and its applications to quantum mechanics of atomic spectra, Academic Press (1959).
2. KAHAN, T., Theory of groups in classical and quantum physics, Oliver and Boyd (1965).
3. MEIJER, P. H. E., Group theory and solid-state physics, Blackie and Son (1964).
4. HAMERMESH, M., Group theory and its applications to physical problems, Pergamon Press (1962).
5. LIUBARSKII, G., Applications of group theory and physics, Pergamon Press (1960).
6. TINKHAM, M., Group theory and quantum mechanics, McGraw-Hill (1963).
7. HEINE, V., Group theory in quantum mechanics, Pergamon (1960).
8. KUROSH, A. G., Theory of groups, Chelsea Publishing Co. (1960).
9. CRACKNELL, A. P., Applied group theory, Pergamon Press (1968).
10. HALL, G. G , Applied group theory, Longmans Green (1967).
11. FALICOV, L. M., Group theory and its physical applications, University of Chicago Press (1966).
12. LOMONT, J. S. Applications of finite groups, Academic Press (1959).

Bibliography

Index

A CATALOG OF SELECTED

DOVER BOOKS

IN SCIENCE AND MATHEMATICS

Astronomy

BURNHAM'S CELESTIAL HANDBOOK, Robert Burnham, Jr. Thorough guide to the stars beyond our solar system. Exhaustive treatment. Alphabetical by constellation: Andromeda to Cetus in Vol. 1; Chamaeleon to Orion in Vol. 2; and Pavo to Vulpecula in Vol. 3. Hundreds of illustrations. Index in Vol. 3. 2,000pp. 6 1/8 x 9 1/4.
Vol. I: 0-486-23567-X
Vol. II: 0-486-23568-8
Vol. III: 0-486-23673-0

EXPLORING THE MOON THROUGH BINOCULARS AND SMALL TELESCOPES, Ernest H. Cherrington, Jr. Informative, profusely illustrated guide to locating and identifying craters, rills, seas, mountains, other lunar features. Newly revised and updated with special section of new photos. Over 100 photos and diagrams. 240pp. 8 1/4 x 11. 0-486-24491-1

THE EXTRATERRESTRIAL LIFE DEBATE, 1750–1900, Michael J. Crowe. First detailed, scholarly study in English of the many ideas that developed from 1750 to 1900 regarding the existence of intelligent extraterrestrial life. Examines ideas of Kant, Herschel, Voltaire, Percival Lowell, many other scientists and thinkers. 16 illustrations. 704pp. 5 3/8 x 8 1/2. 0-486-40675-X

THEORIES OF THE WORLD FROM ANTIQUITY TO THE COPERNICAN REVOLUTION, Michael J. Crowe. Newly revised edition of an accessible, enlightening book re-creates the change from an earth-centered to a sun-centered conception of the solar system. 242pp. 5 3/8 x 8 1/2. 0-486-41444-2

ARISTARCHUS OF SAMOS: The Ancient Copernicus, Sir Thomas Heath. Heath's history of astronomy ranges from Homer and Hesiod to Aristarchus and includes quotes from numerous thinkers, compilers, and scholasticists from Thales and Anaximander through Pythagoras, Plato, Aristotle, and Heraclides. 34 figures. 448pp. 5 3/8 x 8 1/2.
0-486-43886-4

A COMPLETE MANUAL OF AMATEUR ASTRONOMY: TOOLS AND TECHNIQUES FOR ASTRONOMICAL OBSERVATIONS, P. Clay Sherrod with Thomas L. Koed. Concise, highly readable book discusses: selecting, setting up and maintaining a telescope; amateur studies of the sun; lunar topography and occultations; observations of Mars, Jupiter, Saturn, the minor planets and the stars; an introduction to photoelectric photometry; more. 1981 ed. 124 figures. 25 halftones. 37 tables. 335pp. 6 1/2 x 9 1/4. 0-486-42820-8

AMATEUR ASTRONOMER'S HANDBOOK, J. B. Sidgwick. Timeless, comprehensive coverage of telescopes, mirrors, lenses, mountings, telescope drives, micrometers, spectroscopes, more. 189 illustrations. 576pp. 5 5/8 x 8 1/4. (Available in U.S. only.)
0-486-24034-7

STAR LORE: Myths, Legends, and Facts, William Tyler Olcott. Captivating retellings of the origins and histories of ancient star groups include Pegasus, Ursa Major, Pleiades, signs of the zodiac, and other constellations. "Classic."—Sky & Telescope. 58 illustrations. 544pp. 5 3/8 x 8 1/2. 0-486-43581-4

Chemistry

THE SCEPTICAL CHYMIST: THE CLASSIC 1661 TEXT, Robert Boyle. Boyle defines the term "element," asserting that all natural phenomena can be explained by the motion and organization of primary particles. 1911 ed. viii+232pp. $5^3/_8$ x $8^1/_2$.
0-486-42825-7

RADIOACTIVE SUBSTANCES, Marie Curie. Here is the celebrated scientist's doctoral thesis, the prelude to her receipt of the 1903 Nobel Prize. Curie discusses establishing atomic character of radioactivity found in compounds of uranium and thorium; extraction from pitchblende of polonium and radium; isolation of pure radium chloride; determination of atomic weight of radium; plus electric, photographic, luminous, heat, color effects of radioactivity. ii+94pp. $5^3/_8$ x $8^1/_2$. 0-486-42550-9

CHEMICAL MAGIC, Leonard A. Ford. Second Edition, Revised by E. Winston Grundmeier. Over 100 unusual stunts demonstrating cold fire, dust explosions, much more. Text explains scientific principles and stresses safety precautions. 128pp. $5^3/_8$ x $8^1/_2$. 0-486-67628-5

MOLECULAR THEORY OF CAPILLARITY, J. S. Rowlinson and B. Widom. History of surface phenomena offers critical and detailed examination and assessment of modern theories, focusing on statistical mechanics and application of results in mean-field approximation to model systems. 1989 edition. 352pp. $5^3/_8$ x $8^1/_2$. 0-486-42544-4

CHEMICAL AND CATALYTIC REACTION ENGINEERING, James J. Carberry. Designed to offer background for managing chemical reactions, this text examines behavior of chemical reactions and reactors; fluid-fluid and fluid-solid reaction systems; heterogeneous catalysis and catalytic kinetics; more. 1976 edition. 672pp. $6^1/_8$ x $9^1/_4$. 0-486-41736-0 $31.95

ELEMENTS OF CHEMISTRY, Antoine Lavoisier. Monumental classic by founder of modern chemistry in remarkable reprint of rare 1790 Kerr translation. A must for every student of chemistry or the history of science. 539pp. $5^3/_8$ x $8^1/_2$. 0-486-64624-6

MOLECULES AND RADIATION: An Introduction to Modern Molecular Spectroscopy. Second Edition, Jeffrey I. Steinfeld. This unified treatment introduces upper-level undergraduates and graduate students to the concepts and the methods of molecular spectroscopy and applications to quantum electronics, lasers, and related optical phenomena. 1985 edition. 512pp. $5^3/_8$ x $8^1/_2$. 0-486-44152-0

A SHORT HISTORY OF CHEMISTRY, J. R. Partington. Classic exposition explores origins of chemistry, alchemy, early medical chemistry, nature of atmosphere, theory of valency, laws and structure of atomic theory, much more. 428pp. $5^3/_8$ x $8^1/_2$. (Available in U.S. only.) 0-486-65977-1

GENERAL CHEMISTRY, Linus Pauling. Revised 3rd edition of classic first-year text by Nobel laureate. Atomic and molecular structure, quantum mechanics, statistical mechanics, thermodynamics correlated with descriptive chemistry. Problems. 992pp. $5^3/_8$ x $8^1/_2$.
0-486-65622-5

ELECTRON CORRELATION IN MOLECULES, S. Wilson. This text addresses one of theoretical chemistry's central problems. Topics include molecular electronic structure, independent electron models, electron correlation, the linked diagram theorem, and related topics. 1984 edition. 304pp. $5^3/_8$ x $8^1/_2$. 0-486-45879-2

Engineering

DE RE METALLICA, Georgius Agricola. The famous Hoover translation of greatest treatise on technological chemistry, engineering, geology, mining of early modern times (1556). All 289 original woodcuts. 638pp. $6^3/_4$ x 11. 0-486-60006-8

FUNDAMENTALS OF ASTRODYNAMICS, Roger Bate et al. Modern approach developed by U.S. Air Force Academy. Designed as a first course. Problems, exercises. Numerous illustrations. 455pp. $5^3/_8$ x $8^1/_2$. 0-486-60061-0

DYNAMICS OF FLUIDS IN POROUS MEDIA, Jacob Bear. For advanced students of ground water hydrology, soil mechanics and physics, drainage and irrigation engineering and more. 335 illustrations. Exercises, with answers. 784pp. $6^1/_8$ x $9^1/_4$. 0-486-65675-6

THEORY OF VISCOELASTICITY (SECOND EDITION), Richard M. Christensen. Complete consistent description of the linear theory of the viscoelastic behavior of materials. Problem-solving techniques discussed. 1982 edition. 29 figures. xiv+364pp. $6^1/_8$ x $9^1/_4$. 0-486-42880-X

MECHANICS, J. P. Den Hartog. A classic introductory text or refresher. Hundreds of applications and design problems illuminate fundamentals of trusses, loaded beams and cables, etc. 334 answered problems. 462pp. $5^3/_8$ x $8^1/_2$. 0-486-60754-2

MECHANICAL VIBRATIONS, J. P. Den Hartog. Classic textbook offers lucid explanations and illustrative models, applying theories of vibrations to a variety of practical industrial engineering problems. Numerous figures. 233 problems, solutions. Appendix. Index. Preface. 436pp. $5^3/_8$ x $8^1/_2$. 0-486-64785-4

STRENGTH OF MATERIALS, J. P. Den Hartog. Full, clear treatment of basic material (tension, torsion, bending, etc.) plus advanced material on engineering methods, applications. 350 answered problems. 323pp. $5^3/_8$ x $8^1/_2$. 0-486-60755-0

A HISTORY OF MECHANICS, René Dugas. Monumental study of mechanical principles from antiquity to quantum mechanics. Contributions of ancient Greeks, Galileo, Leonardo, Kepler, Lagrange, many others. 671pp. $5^3/_8$ x $8^1/_2$. 0-486-65632-2

STABILITY THEORY AND ITS APPLICATIONS TO STRUCTURAL MECHANICS, Clive L. Dym. Self-contained text focuses on Koiter postbuckling analyses, with mathematical notions of stability of motion. Basing minimum energy principles for static stability upon dynamic concepts of stability of motion, it develops asymptotic buckling and postbuckling analyses from potential energy considerations, with applications to columns, plates, and arches. 1974 ed. 208pp. $5^3/_8$ x $8^1/_2$. 0-486-42541-X

BASIC ELECTRICITY, U.S. Bureau of Naval Personnel. Originally a training course; best nontechnical coverage. Topics include batteries, circuits, conductors, AC and DC, inductance and capacitance, generators, motors, transformers, amplifiers, etc. Many questions with answers. 349 illustrations. 1969 edition. 448pp. $6^1/_2$ x $9^1/_4$. 0-486-20973-3

Mathematics

FUNCTIONAL ANALYSIS (Second Corrected Edition), George Bachman and Lawrence Narici. Excellent treatment of subject geared toward students with background in linear algebra, advanced calculus, physics and engineering. Text covers introduction to inner-product spaces, normed, metric spaces, and topological spaces; complete orthonormal sets, the Hahn-Banach Theorem and its consequences, and many other related subjects. 1966 ed. 544pp. $6^1/_8$ x $9^1/_4$. 0-486-40251-7

DIFFERENTIAL MANIFOLDS, Antoni A. Kosinski. Introductory text for advanced undergraduates and graduate students presents systematic study of the topological structure of smooth manifolds, starting with elements of theory and concluding with method of surgery. 1993 edition. 288pp. $5^3/_8$ x $8^1/_2$. 0-486-46244-7

VECTOR AND TENSOR ANALYSIS WITH APPLICATIONS, A. I. Borisenko and I. E. Tarapov. Concise introduction. Worked-out problems, solutions, exercises. 257pp. $5^5/_8$ x $8^1/_4$. 0-486-63833-2

AN INTRODUCTION TO ORDINARY DIFFERENTIAL EQUATIONS, Earl A. Coddington. A thorough and systematic first course in elementary differential equations for undergraduates in mathematics and science, with many exercises and problems (with answers). Index. 304pp. $5^3/_8$ x $8^1/_2$. 0-486-65942-9

FOURIER SERIES AND ORTHOGONAL FUNCTIONS, Harry F. Davis. An incisive text combining theory and practical example to introduce Fourier series, orthogonal functions and applications of the Fourier method to boundary-value problems. 570 exercises. Answers and notes. 416pp. $5^3/_8$ x $8^1/_2$. 0-486-65973-9

COMPUTABILITY AND UNSOLVABILITY, Martin Davis. Classic graduate-level introduction to theory of computability, usually referred to as theory of recurrent functions. New preface and appendix. 288pp. $5^3/_8$ x $8^1/_2$. 0-486-61471-9

AN INTRODUCTION TO MATHEMATICAL ANALYSIS, Robert A. Rankin. Dealing chiefly with functions of a single real variable, this text by a distinguished educator introduces limits, continuity, differentiability, integration, convergence of infinite series, double series, and infinite products. 1963 edition. 624pp. $5^3/_8$ x $8^1/_2$. 0-486-46251-X

METHODS OF NUMERICAL INTEGRATION (SECOND EDITION), Philip J. Davis and Philip Rabinowitz. Requiring only a background in calculus, this text covers approximate integration over finite and infinite intervals, error analysis, approximate integration in two or more dimensions, and automatic integration. 1984 edition. 624pp. $5^3/_8$ x $8^1/_2$. 0-486-45339-1

INTRODUCTION TO LINEAR ALGEBRA AND DIFFERENTIAL EQUATIONS, John W. Dettman. Excellent text covers complex numbers, determinants, orthonormal bases, Laplace transforms, much more. Exercises with solutions. Undergraduate level. 416pp. $5^3/_8$ x $8^1/_2$. 0-486-65191-6

RIEMANN'S ZETA FUNCTION, H. M. Edwards. Superb, high-level study of landmark 1859 publication entitled "On the Number of Primes Less Than a Given Magnitude" traces developments in mathematical theory that it inspired. xiv+315pp. $5^3/_8$ x $8^1/_2$. 0-486-41740-9

CALCULUS OF VARIATIONS WITH APPLICATIONS, George M. Ewing. Applications-oriented introduction to variational theory develops insight and promotes understanding of specialized books, research papers. Suitable for advanced undergraduate/graduate students as primary, supplementary text. 352pp. $5^{3}/_{8}$ x $8^{1}/_{2}$. 0-486-64856-7

MATHEMATICIAN'S DELIGHT, W. W. Sawyer. "Recommended with confidence" by *The Times Literary Supplement,* this lively survey was written by a renowned teacher. It starts with arithmetic and algebra, gradually proceeding to trigonometry and calculus. 1943 edition. 240pp. $5^{3}/_{8}$ x $8^{1}/_{2}$. 0-486-46240-4

ADVANCED EUCLIDEAN GEOMETRY, Roger A. Johnson. This classic text explores the geometry of the triangle and the circle, concentrating on extensions of Euclidean theory, and examining in detail many relatively recent theorems. 1929 edition. 336pp. $5^{3}/_{8}$ x $8^{1}/_{2}$. 0-486-46237-4

COUNTEREXAMPLES IN ANALYSIS, Bernard R. Gelbaum and John M. H. Olmsted. These counterexamples deal mostly with the part of analysis known as "real variables." The first half covers the real number system, and the second half encompasses higher dimensions. 1962 edition. xxiv+198pp. $5^{3}/_{8}$ x $8^{1}/_{2}$. 0-486-42875-3

CATASTROPHE THEORY FOR SCIENTISTS AND ENGINEERS, Robert Gilmore. Advanced-level treatment describes mathematics of theory grounded in the work of Poincaré, R. Thom, other mathematicians. Also important applications to problems in mathematics, physics, chemistry and engineering. 1981 edition. References. 28 tables. 397 black-and-white illustrations. xvii + 666pp. $6^{1}/_{8}$ x $9^{1}/_{4}$. 0-486-67539-4

COMPLEX VARIABLES: Second Edition, Robert B. Ash and W. P. Novinger. Suitable for advanced undergraduates and graduate students, this newly revised treatment covers Cauchy theorem and its applications, analytic functions, and the prime number theorem. Numerous problems and solutions. 2004 edition. 224pp. $6^{1}/_{2}$ x $9^{1}/_{4}$. 0-486-46250-1

NUMERICAL METHODS FOR SCIENTISTS AND ENGINEERS, Richard Hamming. Classic text stresses frequency approach in coverage of algorithms, polynomial approximation, Fourier approximation, exponential approximation, other topics. Revised and enlarged 2nd edition. 721pp. $5^{3}/_{8}$ x $8^{1}/_{2}$. 0-486-65241-6

INTRODUCTION TO NUMERICAL ANALYSIS (2nd Edition), F. B. Hildebrand. Classic, fundamental treatment covers computation, approximation, interpolation, numerical differentiation and integration, other topics. 150 new problems. 669pp. $5^{3}/_{8}$ x $8^{1}/_{2}$. 0-486-65363-3

MARKOV PROCESSES AND POTENTIAL THEORY, Robert M. Blumental and Ronald K. Getoor. This graduate-level text explores the relationship between Markov processes and potential theory in terms of excessive functions, multiplicative functionals and subprocesses, additive functionals and their potentials, and dual processes. 1968 edition. 320pp. $5^{3}/_{8}$ x $8^{1}/_{2}$. 0-486-46263-3

ABSTRACT SETS AND FINITE ORDINALS: An Introduction to the Study of Set Theory, G. B. Keene. This text unites logical and philosophical aspects of set theory in a manner intelligible to mathematicians without training in formal logic and to logicians without a mathematical background. 1961 edition. 112pp. $5^{3}/_{8}$ x $8^{1}/_{2}$. 0-486-46249-8

Physics

OPTICAL RESONANCE AND TWO-LEVEL ATOMS, L. Allen and J. H. Eberly. Clear, comprehensive introduction to basic principles behind all quantum optical resonance phenomena. 53 illustrations. Preface. Index. 256pp. 5³/₈ x 8¹/₂. 0-486-65533-4

QUANTUM THEORY, David Bohm. This advanced undergraduate-level text presents the quantum theory in terms of qualitative and imaginative concepts, followed by specific applications worked out in mathematical detail. Preface. Index. 655pp. 5³/₈ x 8¹/₂. 0-486-65969-0

ATOMIC PHYSICS (8th EDITION), Max Born. Nobel laureate's lucid treatment of kinetic theory of gases, elementary particles, nuclear atom, wave-corpuscles, atomic structure and spectral lines, much more. Over 40 appendices, bibliography. 495pp. 5³/₈ x 8¹/₂. 0-486-65984-4

A SOPHISTICATE'S PRIMER OF RELATIVITY, P. W. Bridgman. Geared toward readers already acquainted with special relativity, this book transcends the view of theory as a working tool to answer natural questions: What is a frame of reference? What is a "law of nature"? What is the role of the "observer"? Extensive treatment, written in terms accessible to those without a scientific background. 1983 ed. xlviii+172pp. 5³/₈ x 8¹/₂. 0-486-42549-5

AN INTRODUCTION TO HAMILTONIAN OPTICS, H. A. Buchdahl. Detailed account of the Hamiltonian treatment of aberration theory in geometrical optics. Many classes of optical systems defined in terms of the symmetries they possess. Problems with detailed solutions. 1970 edition. xv + 360pp. 5³/₈ x 8¹/₂. 0-486-67597-1

PRIMER OF QUANTUM MECHANICS, Marvin Chester. Introductory text examines the classical quantum bead on a track: its state and representations; operator eigenvalues; harmonic oscillator and bound bead in a symmetric force field; and bead in a spherical shell. Other topics include spin, matrices, and the structure of quantum mechanics; the simplest atom; indistinguishable particles; and stationary-state perturbation theory. 1992 ed. xiv+314pp. 6¹/₈ x 9¹/₄. 0-486-42878-8

LECTURES ON QUANTUM MECHANICS, Paul A. M. Dirac. Four concise, brilliant lectures on mathematical methods in quantum mechanics from Nobel Prize-winning quantum pioneer build on idea of visualizing quantum theory through the use of classical mechanics. 96pp. 5³/₈ x 8¹/₂. 0-486-41713-1

THIRTY YEARS THAT SHOOK PHYSICS: THE STORY OF QUANTUM THEORY, George Gamow. Lucid, accessible introduction to influential theory of energy and matter. Careful explanations of Dirac's anti-particles, Bohr's model of the atom, much more. 12 plates. Numerous drawings. 240pp. 5³/₈ x 8¹/₂. 0-486-24895-X

ELECTRONIC STRUCTURE AND THE PROPERTIES OF SOLIDS: THE PHYSICS OF THE CHEMICAL BOND, Walter A. Harrison. Innovative text offers basic understanding of the electronic structure of covalent and ionic solids, simple metals, transition metals and their compounds. Problems. 1980 edition. 582pp. 6¹/₈ x 9¹/₄. 0-486-66021-4

HYDRODYNAMIC AND HYDROMAGNETIC STABILITY, S. Chandrasekhar. Lucid examination of the Rayleigh-Benard problem; clear coverage of the theory of instabilities causing convection. 704pp. 5⁵/₈ x 8¹/₄. 0-486-64071-X

INVESTIGATIONS ON THE THEORY OF THE BROWNIAN MOVEMENT, Albert Einstein. Five papers (1905–8) investigating dynamics of Brownian motion and evolving elementary theory. Notes by R. Fürth. 122pp. 5³/₈ x 8¹/₂. 0-486-60304-0

THE PHYSICS OF WAVES, William C. Elmore and Mark A. Heald. Unique overview of classical wave theory. Acoustics, optics, electromagnetic radiation, more. Ideal as classroom text or for self-study. Problems. 477pp. 5³/₈ x 8¹/₂. 0-486-64926-1

GRAVITY, George Gamow. Distinguished physicist and teacher takes reader-friendly look at three scientists whose work unlocked many of the mysteries behind the laws of physics: Galileo, Newton, and Einstein. Most of the book focuses on Newton's ideas, with a concluding chapter on post-Einsteinian speculations concerning the relationship between gravity and other physical phenomena. 160pp. 5³/₈ x 8¹/₂. 0-486-42563-0

PHYSICAL PRINCIPLES OF THE QUANTUM THEORY, Werner Heisenberg. Nobel Laureate discusses quantum theory, uncertainty, wave mechanics, work of Dirac, Schroedinger, Compton, Wilson, Einstein, etc. 184pp. 5³/₈ x 8¹/₂. 0-486-60113-7

ATOMIC SPECTRA AND ATOMIC STRUCTURE, Gerhard Herzberg. One of best introductions; especially for specialist in other fields. Treatment is physical rather than mathematical. 80 illustrations. 257pp. 5³/₈ x 8¹/₂. 0-486-60115-3

AN INTRODUCTION TO STATISTICAL THERMODYNAMICS, Terrell L. Hill. Excellent basic text offers wide-ranging coverage of quantum statistical mechanics, systems of interacting molecules, quantum statistics, more. 523pp. 5³/₈ x 8¹/₂. 0-486-65242-4

THEORETICAL PHYSICS, Georg Joos, with Ira M. Freeman. Classic overview covers essential math, mechanics, electromagnetic theory, thermodynamics, quantum mechanics, nuclear physics, other topics. First paperback edition. xxiii + 885pp. 5³/₈ x 8¹/₂. 0-486-65227-0

PROBLEMS AND SOLUTIONS IN QUANTUM CHEMISTRY AND PHYSICS, Charles S. Johnson, Jr. and Lee G. Pedersen. Unusually varied problems, detailed solutions in coverage of quantum mechanics, wave mechanics, angular momentum, molecular spectroscopy, more. 280 problems plus 139 supplementary exercises. 430pp. 6¹/₂ x 9¹/₄. 0-486-65236-X

THEORETICAL SOLID STATE PHYSICS, Vol. 1: Perfect Lattices in Equilibrium; Vol. II: Non-Equilibrium and Disorder, William Jones and Norman H. March. Monumental reference work covers fundamental theory of equilibrium properties of perfect crystalline solids, non-equilibrium properties, defects and disordered systems. Appendices. Problems. Preface. Diagrams. Index. Bibliography. Total of 1,301pp. 5³/₈ x 8¹/₂. Two volumes. Vol. I: 0-486-65015-4 Vol. II: 0-486-65016-2

WHAT IS RELATIVITY? L. D. Landau and G. B. Rumer. Written by a Nobel Prize physicist and his distinguished colleague, this compelling book explains the special theory of relativity to readers with no scientific background, using such familiar objects as trains, rulers, and clocks. 1960 ed. vi+72pp. 5³/₈ x 8¹/₂. 0-486-42806-0

CATALOG OF DOVER BOOKS

A TREATISE ON ELECTRICITY AND MAGNETISM, James Clerk Maxwell. Important foundation work of modern physics. Brings to final form Maxwell's theory of electromagnetism and rigorously derives his general equations of field theory. 1,084pp. $5^3/_8$ x $8^1/_2$. Two-vol. set. Vol. I: 0-486-60636-8 Vol. II: 0-486-60637-6

MATHEMATICS FOR PHYSICISTS, Philippe Dennery and Andre Krzywicki. Superb text provides math needed to understand today's more advanced topics in physics and engineering. Theory of functions of a complex variable, linear vector spaces, much more. Problems. 1967 edition. 400pp. $6^1/_2$ x $9^1/_4$. 0-486-69193-4

INTRODUCTION TO QUANTUM MECHANICS WITH APPLICATIONS TO CHEMISTRY, Linus Pauling & E. Bright Wilson, Jr. Classic undergraduate text by Nobel Prize winner applies quantum mechanics to chemical and physical problems. Numerous tables and figures enhance the text. Chapter bibliographies. Appendices. Index. 468pp. $5^3/_8$ x $8^1/_2$. 0-486-64871-0

METHODS OF THERMODYNAMICS, Howard Reiss. Outstanding text focuses on physical technique of thermodynamics, typical problem areas of understanding, and significance and use of thermodynamic potential. 1965 edition. 238pp. $5^3/_8$ x $8^1/_2$. 0-486-69445-3

THE ELECTROMAGNETIC FIELD, Albert Shadowitz. Comprehensive under- graduate text covers basics of electric and magnetic fields, builds up to electromagnetic theory. Also related topics, including relativity. Over 900 problems. 768pp. $5^3/_8$ x $8^1/_4$. 0-486-65660-8

GREAT EXPERIMENTS IN PHYSICS: FIRSTHAND ACCOUNTS FROM GALILEO TO EINSTEIN, Morris H. Shamos (ed.). 25 crucial discoveries: Newton's laws of motion, Chadwick's study of the neutron, Hertz on electromagnetic waves, more. Original accounts clearly annotated. 370pp. $5^3/_8$ x $8^1/_2$. 0-486-25346-5

EINSTEIN'S LEGACY, Julian Schwinger. A Nobel Laureate relates fascinating story of Einstein and development of relativity theory in well-illustrated, nontechnical volume. Subjects include meaning of time, paradoxes of space travel, gravity and its effect on light, non-Euclidean geometry and curving of space-time, impact of radio astronomy and space-age discoveries, and more. 189 b/w illustrations. xiv+250pp. $8^3/_8$ x $9^1/_4$. 0-486-41974-6

THE VARIATIONAL PRINCIPLES OF MECHANICS, Cornelius Lanczos. Philosophic, less formalistic approach to analytical mechanics offers model of clear, scholarly exposition at graduate level with coverage of basics, calculus of variations, principle of virtual work, equations of motion, more. 418pp. $5^3/_8$ x $8^1/_2$. 0-486-65067-7

⊗
P, 7 $\Theta = \begin{cases} \text{ORIGIN} \\ \text{OBSERVER} \\ \text{ORTHOGONAL} \end{cases}$

⊗
P 9 CONTACT